T0155855

Lecture Notes in Computer Science   12610

More information about this subseries at http://www.springer.com/series/7409

Miao Qiao · Gottfried Vossen ·
Sen Wang · Lei Li (Eds.)

Databases Theory and Applications

32nd Australasian Database Conference, ADC 2021
Dunedin, New Zealand, January 29 – February 5, 2021
Proceedings

Springer

Editors
Miao Qiao 🆔
University of Auckland
Auckland, New Zealand

Gottfried Vossen
University of Münster
Münster, Germany

Sen Wang 🆔
The University of Queensland
St. Lucia, QLD, Australia

Lei Li 🆔
University of Queensland
Brisbane, QLD, Australia

ISSN 0302-9743   ISSN 1611-3349 (electronic)
Lecture Notes in Computer Science
ISBN 978-3-030-69376-3   ISBN 978-3-030-69377-0 (eBook)
https://doi.org/10.1007/978-3-030-69377-0

LNCS Sublibrary: SL3 – Information Systems and Applications, incl. Internet/Web, and HCI

This Springer imprint is published by the registered company Springer Nature Switzerland AG
The registered company address is: Gewerbestrasse 11, 6330 Cham, Switzerland

Preface

It is our pleasure to present to you the proceedings of the 32nd Australasian Database Conference (ADC 2021), which took place in Dunedin, New Zealand. ADC is an annual international forum for sharing the latest research advancements and novel applications of database systems, data-driven applications, and data analytics between researchers and practitioners from around the globe, particularly Australia and New Zealand. The mission of ADC is to share novel research solutions to problems of today's information society that meet the needs of heterogeneous applications and environments, and to identify new issues and directions for future research and development work. ADC seeks papers from academia and industry presenting research on all practical and theoretical aspects of advanced database theory and applications, as well as case studies and implementation experiences. All topics related to databases are of interest and within the scope of the conference. ADC gives researchers and practitioners a unique opportunity to share their perspectives with others interested in the various aspects of database systems.

As in previous years, the ADC 2021 Program Committee accepted papers considered to be of ADC quality without setting any predefined quota. The conference received 21 submissions and accepted 17 full research papers. Each paper was peer reviewed in full by at least three independent reviewers, and in some cases four referees produced independent reviews. A conscious decision was made to select the papers for which all reviews were positive and favorable. The Program Committee that selected the papers consisted of 39 members from around the globe, including Australia, China, Germany, New Zealand, and the USA, who were thorough and dedicated to the reviewing process.

We would like to thank all our colleagues who served on the Program Committee or acted as external reviewers. We would also like to thank all the authors who submitted their papers and the attendees. This conference is held for you, and we hope that with these proceedings, you can have an overview of this vibrant research community and its activities. We encourage you to make submissions to the next ADC conference and contribute to this community.

January 2021

Miao Qiao
Gottfried Vossen
Sen Wang
Lei Li

General Chair's Welcome Message

Tēnā koutou and welcome to the proceedings of the 32nd Australasian Database Conference (ADC 2021). In Australia and New Zealand, ADC is the premier conference on research and applications of database systems, data-driven applications, and data analytics. Over the past decade, ADC has been held in Melbourne (2020), Sydney (2019), Gold Coast (2018), Brisbane (2017), Sydney (2016), Melbourne (2015), Brisbane (2014), Adelaide (2013), Melbourne (2012), and Perth (2011). This year, the ADC conference was part of the Australasian Computer Science Week (ACSW), which was held virtually at Otago University in Dunedin, New Zealand.

After careful consideration by the Programme Committee, a total of 17 research papers were accepted for inclusion in the conference proceedings. We were very fortunate to have four keynote talks presented by world-leading researchers, including Phoebe Chen from La Trobe University, Xuemin Lin from the University of New South Wales, Thomas Lumley from the University of Auckland, and Yanchun Zhang from Victoria University. In addition, we are grateful to Māui Hudson from the University of Waikato, Guodong Long from the University of Technology Sydney, Shirui Pan from Monash University, Bernhard Pfahringer from the University of Waikato, Wei Zhang from the University of Adelaide, and Kaiqi Zhao from the University of Auckland for the invited talks they presented.

I wish to take this opportunity to thank all the speakers, authors, and organizers. My special gratitude goes out to the Program Committee Co-chairs Miao Qiao, Gottfried Vossen, and Sen Wang for their dedication in ensuring a high-quality program, all members of the Programme Committee for their commitment in providing high-quality reviews, Publication Chair Lei Li for his timely preparation of the conference proceedings, the organizers of ACSW for making it possible to run all ACSW conferences virtually, and Publicity Chair Weitong Chen, for his efforts in disseminating our call for papers and attracting submissions. Without them, this year's ADC would not have been a success.

Dunedin is the second-largest city on the South Island of New Zealand, and famous for being a centre of tertiary education. We hope all contributors had a wonderful experience with the conference. We look forward to welcoming any participant of ACSW 2021 to the beautiful country of Aotearoa in the near future.

Sebastian Link

Organization

General Chair

Sebastian Link   The University of Auckland, New Zealand

Program Committee Chairs

Miao Qiao   The University of Auckland, New Zealand
Gottfried Vossen   University of Münster, Germany
Sen Wang   The University of Queensland, Australia

Publication Chair

Lei Li   The University of Queensland, Australia

Steering Committee

Rao Kotagiri   The University of Melbourne, Australia
Timos Sellis   Swinburne University of Technology, Australia,
     and Facebook, USA
Gill Dobbie   The University of Auckland, New Zealand
Alan Fekete   The University of Sydney, Australia
Xuemin Lin   University of New South Wales, Australia
Yanchun Zhang   Victoria University, Australia
Xiaofang Zhou   Hong Kong University of Science and Technology,
     Hong Kong SAR, China

Program Committee

Ash Rahimi   University of Queensland, Australia
Dan He   University of Queensland, Australia
Farhana Choudhury   University of Melbourne, Australia
Goce Ristanoski   University of Melbourne, Australia
Gottfried Vossen   University of Münster, Germany
Guangyan Huang   Deakin University, Australia
Guodong Long   University of Technology Sydney, Australia
Janusz Getta   University of Wollongong, Australia
Jiangzhang Gan   Massey University, New Zealand
Jianxin Li   Deakin University, Australia
Jinli Cao   La Trobe University, Australia
Junhao Gan   University of Melbourne, Australia
Junhu Wang   Griffith University, Australia

Contents

Intention Recognition from Spatio-Temporal Representation of EEG Signals

Lin Yue[1], Dongyuan Tian[2], Jing Jiang[3], Lina Yao[4], Weitong Chen[5], and Xiaowei Zhao[1(✉)]

[1] Northeast Normal University, Changchun, China
zhaoxw303@nenu.edu.cn
[2] Jilin University, Changchun, China
[3] The University of Technology Sydney, Sydney, Australia
[4] University of New South Wales, Sydney, Australia
[5] The University of Queensland, Brisbane, Australia

Abstract. The motor imagery brain-computer interface uses the human brain intention to achieve better control. The main technical problems are feature representation and classification of signal features for specific thinking activities. Inspired by the structure and function of the human brain, we construct a neural computing model to explore the critical issues in the representation and real-time recognition of the state of specific thinking activities. In consideration of the physiological structure and the information processing process of the brain, we construct a multi-scale cascaded Conv-GRU model and extract high-resolution feature information from the dual spatio-temporal dimension, effectively removing signal noise, improving the signal-to-noise ratio, and reducing information loss. Extensive experiments demonstrate that our model has a low dependence on training data size and outperforms state-of-the-art multi-intention recognition methods.

Keywords: Brain-computer interface · Motor imagery · Electroencephalography · Intention recognition

1 Introduction

Brain-computer interface (BCI) can convert neuron activities into signals, thus providing the possibility for discovering the correlation between brain activities and human behaviors. The electroencephalography (EEG) collected by BCI records brain activities with electrophysiological indicators. During brain activity, the sum of postsynaptic potentials is generated synchronously by a large number of neurons. This process records the electrical wave changes during brain activity, reflecting the electrophysiological activities of brain nerve cells in the cerebral cortex or on scalp surface. By analyzing and modeling EEG signals, such models could be applied to clinical practice such as EEG signal-controlled

© Springer Nature Switzerland AG 2021
M. Qiao et al. (Eds.): ADC 2021, LNCS 12610, pp. 1–12, 2021.
https://doi.org/10.1007/978-3-030-69377-0_1

wheelchairs [8], brain wavelet-controlled exoskeleton [9], brain-controlled hearing aid [10], biomedical implant antennas [1], and motor function recovery during rehabilitation [6]. Other application fields include smart living [27], and speech synthesis [3], etc. Through BCI technology, external devices can read brain nerve signals and convert thinking activities into command signals to realize the human mind control. As a result, EEG based intention recognition has been widely studied in recent years and has become one of the most important research topics in pattern recognition.

1.1 Motivation

In EEG signal analysis, the EEG signal segment includes different frequency bands, each with different degrees of correlation with specific brain activity. Specifically, the frequency band represents brain state and qualitative assessment of awareness; the whole band is between 0.5 Hz to 28 Hz [12]. This interval signal can be decomposed into six types of waves, i.e., Delta, Theta, Alpha, Beta1, Beta2, and Beta3. These waves record the characteristics of the motor or sensory nerve action potentials. Among them, the Alpha wave fluctuates in the state of eyes closed and relaxation, while the Beta wave is closely related to motion behavior and attenuation of motion [24]. Different noise levels distribute in these frequency bands, requiring to be removed via adequate measures. Using and separating multiple waves can help to capture correlations between waves and significant features [13,15,16]. On the other hand, extracting the correlations of temporal and spatial features in all signal bands will improve the performance of intention recognition [25]. On this basis, data flow visualization can help to better understand the whole process of brain activity [4]. However, the factors mentioned above have not been fully taken into account when performing motion intention recognition.

1.2 Challenges

EEG based intention recognition developed rapidly and achieved specific gratifying results. Nevertheless, due to technical limitations, there are still some challenges:

- The brain signals are easily disrupted by a variety of biological signals and environmental artifacts.
- Due to the non-stationary characteristics of electrophysiological brain signals, the raw EEG signals have a low signal-to-noise ratio (SNR).
- Existing machine learning studies focus on static data, so it is impossible to classify rapidly changing brain signals accurately.

1.3 Solution

To overcome these challenges, we take temporal features in multivariate time series and spatial information into our consideration during feature representation and develop a model for multi-intention recognition (Fig. 1). To be more specific, we decompose mixed EEG signal collected with each electrode into signals at frequency bands to reduce noise caused by other frequency ranges and slice the signal series of multiple electrodes into matrices with a sliding window. For filtered EEG arrays, the image mapping layer is utilized for processing EEG arrays into visual images. Finally, we propose multi-scale cascaded Conv-GRU networks (MCG) for image learning with spatio-temporal information. The architecture of the neural network consists of another two parts, i.e., cascaded CNN (convolutional neural network) and GRU (gated recurrent unit), which are used to learn spatio-temporal characteristics, respectively. The multi-intention recognition of dynamic data streams can be effectively solved in this way.

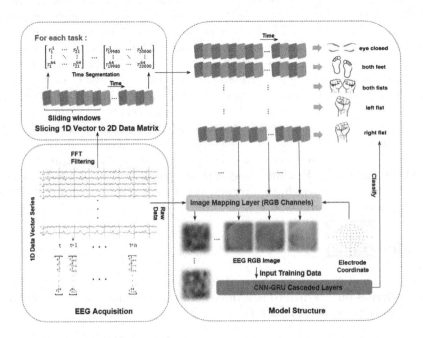

Fig. 1. The workflow of the proposed model MCG.

2   Related Work

Intention recognition can be treated as a classification problem, predicting multiple and subjective human intentions based on EEG traces, rather than actions triggered by events or environment.

Deep learning has been successfully applied in many recognition tasks corresponding to various types of data such as image, video, speech, and text [11,20]. These methods can also be migrated to the task of EEG signal detection. For example, Alomari et al. [2] use a wireless EEG headset as a remote control for a personal computer's mouse cursor. Moreover, in their method, SVM is used for a binary classification task. Kim et al. [13] obtain the Mu and Beta rhythms from the nonlinear EEG signals and perform prediction using a random forest classifier. Zhang et al. [26] apply deep recurrent neural networks on EEG data and improve performance in multiple classification tasks. At present, the classic solution and EEG state recognition technology are respectively used to select features of continuous-time series and distinguish manifolds between learning states through supervised learning [14]. In BCI systems, processing dynamic data flows often require feature representation with spatio-temporal clues. Modeling the correlations between EEG wavelets and multiple intentions and the problem of multi-intention recognition on dynamical data streams are not well solved yet.

A lot of research has adopted CNN for classification on single-channel EEG [21,23]. At the same time, the feature mapping needs to communicate the complexity of the information without losing original richness or depth. Many successful cases apply ConvNets (convolutional neural networks) to distinguish pathological records from normal EEG recordings in the Temple University Hospital EEG Abnormal Corpus. Furthermore, visualization of the ConvNet decoding behavior shows that they use spectral power changes in Delta (0–4 Hz) and Theta (4–8 Hz) frequency ranges [17]. Similar work has been performed based on deep ConvNets to improve decoding errors in EEG signals of human observers [5]. Among the visual researches of high-dimensional EEG data, many methods visualize data as snapshots or sequential images showing the changing trend by time-lapse method [4]. However, most of the mapping methods in these work use more types of waves, which, to a certain extent, increase the complexity of the method and waste more resources.

3   Method

This section will describe the proposed model multi-scale cascaded Conv-GRU in detail, which is further divided into three parts: data acquisition, image mapping layer, and architecture of neural network.

3.1   Data Acquisition

The EEG signals in this paper are based on the BCI system and collected with a 64/14 electrodes headset. The design of data pretreatment is based on the EEG source data. Specifically, once the subject's action command is given, the 64/14 electrodes will pick up brain signals that reflect the brain activities of different areas. Once a subject generates an intention in mind, the electrodes will pick up voltage fluctuations that reflect multiple brain activities. The voltage values from the scalp will be continuously captured by 64-channel or 14-channel electrode sensors. EEG reading can be represented with a n-dimensional vector $R_t = [r_t^1, r_t^2, ..., r_t^n]$, where the r_t^i is the reading of ith electrode sensor at time step t, it can be seen as 1D vector with a certain amount of noise.

It is commonly known that EEG signals can also be divided into multiple data streams according to frequency ranges, with each band having biological significance [9, 22]. The EEG signals consist of multiple time series corresponding to the measurements at different frequency bands [7]. The EEG signal can be quantified in the frequency range from 0.5 Hz to 28 Hz [5]. The raw EEG signal in r_t^i can be segmented into different categories of bandwidth c, where $c = (\delta, \theta, \alpha, \beta)$. This study focuses on two frequency bands from 8 Hz to 28 Hz, i.e., α and β.

Next, we will define the sliding window that further divides the filtered data. To begins with, we need to ensure the maximum value of the possible window scale. As the data in the EEGMMIDB dataset was collected from 64 electrodes, the sliding window dimension is set as [64, 1] with sliding step size 1 here. The data slices are generated along the time axis, and the resulting data matrix is named a sliding matrix here. Finally, we can get N matrices with spatio-temporal characteristics from the raw data in this way. The data segment is created as follows:

$$S = [s_t, s_{t+1}, ..., s_{t+N-1}] \tag{1}$$

$$s_t = \begin{bmatrix} r_t^1 \\ r_t^2 \\ \vdots \\ r_t^{64} \end{bmatrix} \approx \begin{bmatrix} r_t^{\alpha,1} & r_t^{\beta,1} & r_t^{raw,1} \\ r_t^{\alpha,2} & r_t^{\beta,2} & r_t^{raw,2} \\ \vdots & \vdots & \vdots \\ r_t^{\alpha,64} & r_t^{\beta,64} & r_t^{raw,64} \end{bmatrix} \tag{2}$$

where S_j is the jth data segment at time step $t + j - 1$, $\forall j \in [1...N]$; each electrode r_t^i corresponds to three readings (say $[r_t^{\alpha,i}, r_t^{\beta,i}, r_t^{raw,i}]$) at time t, as the filtering operation is adopted.

3.2   Image Mapping Layer

We convert the spatial distribution of electrodes in three-dimensional space into coordinates in two-dimensional space while preserving the relative distance between adjacent electrodes, as shown in Fig. 2. Specifically, this two-dimensional space is a 32×32 mesh, where each pixel in the mesh is superimposed by three channels, and each channel corresponds to a selected frequency band [4]. In this paper, we select α, β, and raw data as the input of three channels. The next step is the normalization that constrains the data range in a closed interval $[0, 255]$. The image synthesizer combines regularized data and generates suitable pixel values for each targeting coordinate. As different electrodes represent different brain regions, the real-time viewer can capture dynamic results in real-time. The width and height of the energy map represent the spatial distribution of mind activities in the cerebral cortex, and the energy map sequence represents the temporal distribution of mind activities. The energy map sequence (image sequence) $NRGMapSeq$ can be denoted as follows:

$$NRGMapSeq = [I_t, I_{t+1}, ..., I_{t+N-1}] \tag{3}$$

where $NRGMapSeq_j$ is the jth energy map (image) at time step $t + j - 1$, $\forall j \in [1...N]$.

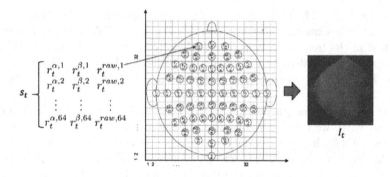

Fig. 2. The process of EEG data segment to image.

3.3 Architecture of the Neural Network

The input of the model is the sequence of 3-channel images (energy maps), which represent the spatial-temporal EEG information. Firstly, the cascaded convolutional neural network will catch partial distribution features from the fragments in image sequence. The performance of the model enhances as the number of convolutional cascade layers increase. After that, the GRU will receive a vector of time series processed by the convolutional cascade layer and further optimize time feature learning.

In order to get detailed and sufficient spatial distribution, the input images can be expressed as:

$$NRGMapSeq = [I_t, I_{t+1}...I_{t+N-1}] \in \mathbb{R}^{N \times c \times h \times w} \tag{4}$$

where N denotes the number of energy maps (images) and the size of each energy map (image) is $c \times h \times w$ (3 channels, height of 32 pixels, width of 32 pixels).

The energy map (image) sequence is input into a Conv2D (two-dimensional convolutional neural networks), and each of the spatial features extracted from the cascaded Conv2D representation is shown in Eq. (5).

$$SP = C_{conv2D}(NRGMapSeq) \tag{5}$$

After the cascaded Conv2D layer, a fully connected layer is applied to connect cascaded Conv2D with the next GRU layer. The RNN has sufficient ability to process arbitrary sequential inputs by recursively applying a transition function to hidden vector h_t. The activation function of the current hidden state h_t at t time step can be computed as follows:

$$h_t = \begin{cases} 0 & t = 0 \\ \int (h_{t-1}, x_t) & \text{otherwise} \end{cases} \tag{6}$$

where x_t is the current state input, and h_{t-1} is the previous hidden state. However, RNN has difficulty learning long-term dependency. The components of the gradient vector will vanish or explode exponentially over a long sequence. As a variant of the LSTM (long short-term memory), the GRU synthesizes the forget gate and input gate into one single update gate. Moreover, there is also a mixture of cellular and hidden states with other modifications. The final model is more straightforward than the standard LSTM model. Both LSTM and GRU can retain important features through various gates to ensure that they will not be lost in long-term propagation. Moreover, the GRU transition equations are defined as follows:

$$z_t = \sigma(W_z \cdot [h_{t-1}, x_t])$$
$$r_t = \sigma(W_r \cdot [h_{t-1}, x_t]) \tag{7}$$
$$h_t = (1 - z_t) \times h_{t-1} + z_t \times \tanh(W \cdot [r_t \times h_{t-1}, x_t])$$

4 Experiments

4.1 Datasets

To verify the validity of the proposed method, we tested the proposed method and all the benchmarking methods with cross-validation on EEGMMIDB[1] and EMOTIV[2], respectively. The intention recognition is treated as a classification task; that is to say, the proposed method MCG will classify five types of intention for both datasets.

4.2 Benchmarking Methods

We compared the proposed model against various state-of-the-art methods. For the baseline models, we kept the same structures and settings. We fed baselines with different features extracted from the same datasets to evaluate the influence of multi-resolution signals. Moreover, a brief introduction of the benchmarking methods as described below:

- Alomari et al. [2]: A support vector machine-based method is used for binary classification, along with features extracted from multi-resolution EEG signals.
- Shenoy et al. [18]: Regularisation is deployed to improve the robustness and accuracy of CSP estimation in features extracting processing. Fisher linear discriminant is used to perform binary tasks.
- Rashid et al. [16]: Neural network (NN) is utilized to perform EEG signal binary-class tasks after decomposing the raw EEG data to extract significant features.
- Kim et al. [13]: Random forest classifier is used for prediction, in which the Mu and Beta rhythms are obtained from the nonlinear EEG signals.
- Sita et al. [19]: Features are extracted from open source EEG data, and LDA solves multiple classification problems.

---

[1] https://physionet.org/pn4/eegmmidb/.
[2] https://drive.google.com/drive/folders/0B9MuJb6Xx2PIM0otakxuVHpkWkk.

– Zhang et al. [26]: Deep recurrent neural networks are applied on an open EEG database for multiple classifications.
– Chen et al. [7]: Multi-task RNNs model (MTLEEG) is proposed for motion intention recognition based EEG signals.

4.3   Results and Discussion

Visual Verification and Analysis. The image mapping layer generates the brain energy maps, in which each image represents the spatial distribution of the corresponding areas for mind activities, and the energy map sequence represents the temporal distribution or dynamic real-time results. From Fig. 3(a), we can intuitively observe the energy changes corresponding to different actions on different imagery tasks. As displayed in Fig. 3(b), we reserved two brain energy mapping channel respectively. In this way, we can clearly understand the spatio-temporal characteristics of both the two waveforms, which provides more possibilities and ideas for brain working mechanism research.

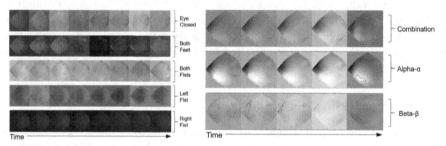

(a) Filtered EEG (Channel 1: α; Channel (b) Decomposition of waveforms. (Channel 1: α; 2: β; Channel 3: Raw)   Channel 2: β; Channel 3: -)

Fig. 3. Brain energy maps from the image mapping layer on EEGMMIDB dataset.

Table 1. Comparisons of different waveband components on EEGMMIDB.

Method	Accuracy	Precision	Recall	F1-score	AUC
MCG-α	0.7722	0.8763	0.7121	0.7857	0.9500
MCG-β	0.8923	0.7707	0.8144	0.7919	0.9619
MCG-α, β	0.9650	0.9650	**0.9806**	0.9727	0.9740
MCG	**0.9870**	**0.9981**	0.9681	**0.9829**	**0.9740**

Effect of Filtering. In this subsection, we used different combinations of waveforms to test the model in terms of accuracy, precision, recall, F1-score, and AUC. As shown in Table 1, for the combination of input signal (α, β, raw), MCG achieves the best performance on multiple indicators on the whole, which directly verifies the importance of filtering and the combination of raw signal

Table 2. Comparisons of evolution models on EEGMMIDB.

Input	Method	Class	Accuracy
α	CNN	Multiple (5)	0.7109
α	CasCNN	Multiple (5)	0.7556
α	CasCNN+GRU	Multiple (5)	**0.7722**
β	CNN	Multiple (5)	0.7161
β	CasCNN	Multiple (5)	0.7632
β	CasCNN+GRU	Multiple (5)	**0.8923**
α, β	CNN	Multiple (5)	0.8218
α, β	CasCNN	Multiple (5)	0.7723
α, β	CasCNN+GRU	Multiple (5)	**0.9829**
α, β, raw	CNN	Multiple (5)	0.8401
α, β, raw	CasCNN	Multiple (5)	0.9356
α, β, raw	**CasCNN+GRU**	Multiple (5)	**0.9868**

with α and β. This combination maximally preserves the useful wavebands and avoids the loss of the original features of the data.

Comparisons of Evolution Models. Furthermore, we compared MCG with evolution models of CNN and cascade structure CNN (see Table 2). In the scope of this paper, the evolution model means we remove some layers in the model, such as GRU and CasCNN, and only use the deep convolutional network (CNN) to train the data. By observing the influence of these layers on the experimental results, we could see that both GRU and CasCNN improved the accuracy.

Comparisons of MCG and Benchmarking Methods. To prove the generalization and robustness of MCG, we further compared MCG with multiple state-of-the-art methods, on EEGMMIDB and EMOTIV datasets. Although the EMOTIV is collected with EMOTIV Epoc+ headset, which contains fewer sensors and has a lower sampling rate, i.e., 14 sensors and 128 Hz sampling rate. The comparisons as shown in Table 3 and 4, have vividly illustrated that MCG achieve stable and brilliant performance in terms of accuracy.

Table 3. Comparisons of MCG and benchmarking methods on EEGMMIDB.

Index	Method	Class	Accuracy
1	Almoari et al. [2]	Binary	0.7500
2	Shenoy et al. [18]	Binary	0.8206
3	Rashid et al. [16]	Binary	0.9199
4	Kim et al. [13]	Multiple (3)	0.8050
5	Sita et al. [19]	Multiple (3)	0.8500
6	Zhang et al. [26]	Multiple (5)	0.9590
7	Chen et al. [7]	Multiple (5)	0.9786
8	**MCG**	Multiple (5)	**0.9868**

Table 4. Comparisons of MCG and benchmarking methods on EMOTIV.

Index	Method	Class	Accuracy
1	Almoari et al. [2]	Binary	0.5627
2	Shenoy et al. [18]	Binary	0.5553
3	Rashid et al. [16]	Binary	0.7538
4	Kim et al. [13]	Multiple (3)	0.7695
5	Sita et al. [19]	Multiple (3)	0.6985
6	Zhang et al. [26]	Multiple (5)	0.7361
7	Chen et al. [7]	Multiple (5)	0.8396
8	**MCG**	Multiple (5)	**0.8600**

5   Conclusions

In this paper, we propose MCG model, which uses the image mapping layer to capture spatial information of the EEG signals and combines spatial-temporal characteristics to identify multiple motion intentions. The proposed model is capable of discovering the brain changes corresponding to different actions. That is, the feature representation achieved through the image mapping layer reflects not only the changes in brain-related different movements but also dynamic responses of the brain in real-time corresponding to specific actions. Experimental results illustrate that the recognition efficiency is the highest among state-of-the-are methods on the multi-classification task of intention recognition.

Acknowledgement. This research has been supported by the Fundamental Research Funds for the Central Universities under Grant No. 2412019FZ047, the China Postdoctoral Science Foundation under Grant No. 2017M621192, the National Natural Science Foundation of China (NSFC) under Grant No.61972384, and the Outstanding Sino-foreign Youth Exchange Program of China Association for Science and Technology.

References

1. Agarwal, K., Guo, Y.X.: Interaction of electromagnetic waves with humans in wearable and biomedical implant antennas. In: 2015 Asia-Pacific Symposium on Electromagnetic Compatibility (APEMC), pp. 154–157. IEEE (2015)
2. Alomari, M.H., AbuBaker, A., Turani, A., Baniyounes, A.M., Manasreh, A.: EEG mouse: a machine learning-based brain computer interface. Int. J. Adv. Comput. Sci. Appl. **5**(4), 193–198 (2014)
3. Anumanchipalli, G.K., Chartier, J., Chang, E.F.: Speech synthesis from neural decoding of spoken sentences. Nature **568**(7753), 493 (2019)
4. Bashivan, P., Rish, I., Yeasin, M., Codella, N.: Learning representations from EEG with deep recurrent-convolutional neural networks. arXiv preprint arXiv:1511.06448 (2015)

5. Behncke, J., Schirrmeister, R.T., Burgard, W., Ball, T.: The signature of robot action success in EEG signals of a human observer: Decoding and visualization using deep convolutional neural networks. In: 2018 6th International Conference on Brain-Computer Interface (BCI), pp. 1–6. IEEE (2018)
6. Biryukova, E., et al.: Arm motor function recovery during rehabilitation with the use of hand exoskeleton controlled by brain-computer interface: a patient with severe brain damage. Fiziol. Cheloveka **42**(1), 19–30 (2016)
7. Chen, W., et al.: EEG-based motion intention recognition via multi-task RNNs. In: Proceedings of the 2018 SIAM International Conference on Data Mining, pp. 279–287. SIAM (2018)
8. Fiala, P., Hanzelka, M., Čáp, M.: Electromagnetic waves and mental synchronization of humans in a large crowd. In: 2017 11th International Conference on Measurement, pp. 241–244. IEEE (2017)
9. Frolov, A.A., Húsek, D., Biryukova, E.V., Bobrov, P.D., Mokienko, O.A., Alexandrov, A.: Principles of motor recovery in post-stroke patients using hand exoskeleton controlled by the brain-computer interface based on motor imagery. Neural Netw. World **27**(1), 107 (2017)
10. Han, C., O'Sullivan, J., Luo, Y., Herrero, J., Mehta, A.D., Mesgarani, N.: Speaker-independent auditory attention decoding without access to clean speech sources. Sci. Adv. **5**(5), eaav6134 (2019)
11. He, K., Zhang, X., Ren, S., Sun, J.: Spatial pyramid pooling in deep convolutional networks for visual recognition. IEEE Trans. Pattern Anal. Mach. Intell. **37**(9), 1904–1916 (2015)
12. Kaiser, A.K., Doppelmayr, M., Iglseder, B.: EEG beta 2 power as surrogate marker for memory impairment: a pilot study. Int. Psychogeriatr. **29**(9), 1515–1523 (2017)
13. Kim, Y., Ryu, J., Kim, K.K., Took, C.C., Mandic, D.P., Park, C.: Motor imagery classification using mu and beta rhythms of EEG with strong uncorrelating transform based complex common spatial patterns. Comput. Intell. Neurosci. **2016**, 1 (2016)
14. Lotte, F., Congedo, M., Lécuyer, A., Lamarche, F., Arnaldi, B.: A review of classification algorithms for EEG-based brain-computer interfaces. J. Neural Eng. **4**(2), R1 (2007)
15. Moore, M.R., Franz, E.A.: Mu rhythm suppression is associated with the classification of emotion in faces. Cogn. Affect. Behav. Neurosci. **17**(1), 224–234 (2016). https://doi.org/10.3758/s13415-016-0476-6
16. or Rashid, M.M., Ahmad, M.: Classification of motor imagery hands movement using Levenberg-Marquardt algorithm based on statistical features of EEG signal. In: 2016 3rd International Conference on Electrical Engineering and Information Communication Technology (ICEEICT), pp. 1–6. IEEE (2016)
17. Schirrmeister, R.T., et al.: Deep learning with convolutional neural networks for EEG decoding and visualization. Hum. Brain Mapp. **38**(11), 5391–5420 (2017)
18. Shenoy, H.V., Vinod, A.P., Guan, C.: Shrinkage estimator based regularization for EEG motor imagery classification. In: 2015 10th International Conference on Information, Communications and Signal Processing (ICICS), pp. 1–5. IEEE (2015)
19. Sita, J., Nair, G.: Feature extraction and classification of EEG signals for mapping motor area of the brain. In: 2013 International Conference on Control Communication and Computing (ICCC), pp. 463–468. IEEE (2013)
20. Song, S., Miao, Z.: Research on vehicle type classification based on spatial pyramid representation and BP neural network. In: Zhang, Y.-J. (ed.) ICIG 2015. LNCS, vol. 9219, pp. 188–196. Springer, Cham (2015). https://doi.org/10.1007/978-3-319-21969-1_17

21. Sors, A., Bonnet, S., Mirek, S., Vercueil, L., Payen, J.F.: A convolutional neural network for sleep stage scoring from raw single-channel EEG. Biomed. Signal Process. Control **42**, 107–114 (2018)
22. Tatum, W.O.: Ellen R. grass lecture: extraordinary EEG. Neurodiagnostic J. **54**(1), 3–21 (2014)
23. Tsinalis, O., Matthews, P.M., Guo, Y., Zafeiriou, S.: Automatic sleep stage scoring with single-channel EEG using convolutional neural networks. arXiv preprint arXiv:1610.01683 (2016)
24. Wang, S., Chang, X., Li, X., Long, G., Yao, L., Sheng, Q.Z.: Diagnosis code assignment using sparsity-based disease correlation embedding. IEEE Trans. Knowl. Data Eng. **28**(12), 3191–3202 (2016)
25. Zhang, D., Yao, L., Zhang, X., Wang, S., Chen, W., Boots, R.: EEG-based intention recognition from spatio-temporal representations via cascade and parallel convolutional recurrent neural networks. arXiv preprint arXiv:1708.06578 (2017)
26. Zhang, X., Yao, L., Huang, C., Sheng, Q.Z., Wang, X.: Intent recognition in smart living through deep recurrent neural networks. In: Liu, D., Xie, S., Li, Y., Zhao, D., El-Alfy, E.S. (eds.) ICONIP 2017. LNCS, vol. 10635, pp. 748–758. Springer, Cham (2017). https://doi.org/10.1007/978-3-319-70096-0_76
27. Zhang, X., Yao, L., Sheng, Q.Z., Kanhere, S.S., Gu, T., Zhang, D.: Converting your thoughts to texts: enabling brain typing via deep feature learning of EEG signals. In: 2018 IEEE International Conference on Pervasive Computing and Communications (PerCom), pp. 1–10. IEEE (2018)

Adaptive Graph Learning for Semi-supervised Classification of GCNs

Yingying Wan[1], Mengmeng Zhan[1], and Yangding Li[2(✉)]

[1] Guangxi Key Lab of Multi-source Information Mining and Security, Guangxi Normal University, Guilin, Guangxi, China
[2] Hunan Provincial Key Laboratory of Intelligent Computing and Language Information Processing, Hunan Normal University, Changsha, China
lyd_175@163.com

Abstract. Graph convolutional networks (GCNs) have achieved great success in social networks and other aspects. However, existing GCN methods generally require a wealth of domain knowledge to obtain the data graph, which cannot guarantee that the graph is suitable. In this paper, we propose adaptive graph learning for semi-supervised classification of GCNs. Firstly, the hypergraph is used to establish the initial neighborhood relationship between data. Then hypergraph, sparse learning and adaptive graph are integrated into a framework. Finally, the suitable graph is obtained, which is inputted into GCN for semi-supervised learning. The experimental results of multi-type datasets show that our method is superior to other comparison algorithms in classification tasks.

Keywords: Graph convolutional networks · Adaptive graph learning · Hypergraph · Laplace

1 Introduction

Graphs, as the effective representations of data distribution, play an important role in describing the inherent structure of data [11,22]. Many learning tasks in the real world can be described as graph problems [21,25]. For example, in the field of biotechnology, graphs can be used to describe the internal structure of protein [8]. In the field of social networks, graph structures are used to describe the relationship among many people or groups [2]. Recently, researchers try to use graph convolutional networks (GCNs) to deal with graph data [17,19] in deep learning.

This work is supported in part by the National Natural Science Foundation of China (Grant No: 81701780); the Guangxi Natural Science Foundation (Grant No: 2017GXNSFBA198221); the Project of Guangxi Science and Technology (GuiKeAD19110133, GuiKeAD20159041); the Innovation Project of Guangxi Graduate Education (Grants No: YCSW20201008, JXXYYJSCXXM-008); and the Hunan Provincial Science & Technology Project Foundation (2018TP1018, 2018RS3065).

M. Qiao et al. (Eds.): ADC 2021, LNCS 12610, pp. 13–22, 2021.
https://doi.org/10.1007/978-3-030-69377-0_2

Generally speaking, there are two kinds of GCN methods, spectral convolution methods [6,13] and spatial convolution methods [1]. The spectral convolution methods are generally defined as the convolution operation based on the spectral representation of the graph. Defferrard et al. [4] used Chebyshev polynomials as the convolution kernel to reduce the computational complexity. Yadati et al. [26] proposed HyperGCN, which used a layered propagation rule for convolutional networks to operate directly on hypergraphs. Compared with convolution in spectral domain, the spatial convolution methods have better expansibility. For example, Atwood et al. [1] proposed diffusion graph convolution network (DCNN), which can be used to model graph classification. Veličković et al. [24] proposed Graph Attention Network (GAT) with designing an attention layer.

The above methods have been widely used in supervised or semi-supervised graph convolutional networks [29,30]. One important aspect of GCNs is the graph representation of data, which is easily ignored [12,31]. In general, the data provided to GCNs is a known intrinsic graph structure or artificially constructed a graph [9]. However, these graph methods are easy to be affected by impurities (such as noise or redundancy) contained in the data, and the quality of the constructed graph is usually difficult to be guaranteed [2,10].

So we propose adaptive graph learning for semi-supervised classification of GCNs. Considering the influence of noise on graph structure, we introduce hypergraph to represent the initial relationship between data samples. Then adaptive graph Laplacian learning and sparse learning are integrated into a framework to get the suitable graph and input it into GCN for semi-supervised learning. Experimental results show that our proposed method has better robustness and superiority than other comparison methods in different data sets. Our method has main advantages:

– We use hypergraph to define the initial graph, which fully considers the multiple relationships of data.
– We introduce adaptive graph learning and sparse learning to get a more suitable graph for semi-supervised classification of GCNs.

2   Related Work

We have the following definitions in this paper. Matrices, vectors and scalars are represented as boldface uppercase letters, bold lowercase letters and normal italic letters, respectively. For a matrix $\mathbf{M} = \{\mathbf{m}_i\}_{i=1}^n \in \mathbb{R}^{n \times d}$, the i-th row and the i, j-th of $\mathbf{M}$ are denoted as $\mathbf{m}_i$ and m_{ij}. We denote the F-norm of $\mathbf{M}$ as $\|\mathbf{M}\|_F = \sqrt{\sum_{ij} |m_{ij}|^2}$. Besides, we denote $\mathbf{M}^T$ and $tr(\mathbf{M})$ as the transpose of $\mathbf{M}$ and the trace of $\mathbf{M}$, respectively.

2.1   Hypergraph Theory

Hypergraph $\mathbf{G}\,(\mathbf{V}, \mathbf{E}, \mathbf{M})$ is composed of a set of vertices $\mathbf{V}$, a set of hyperedges $\mathbf{E}$, and a matrix of hyperedge weights $\mathbf{M}$. Each hyperedge e_i is defined with a weight m_i. The link relation of the hypergraph $\mathbf{G}$ is expressed as $\mathbf{H} \in {}^{|V| \times |E|}$, the correlation matrix $\mathbf{H}$ defined as follows:

$$\mathbf{H}(v, e) = \begin{cases} 1, if v \in e, \\ 0, if v \notin e. \end{cases} \tag{1}$$

According to the literature [3,6], we define the adjacency matrix of a standard hypergraph $\mathbf{A}$:

$$\mathbf{A} = \mathbf{HMH}^T - \mathbf{D}_v, \tag{2}$$

where $\mathbf{D}_v$ is the angle matrix of each vertex [16]. In the literature [23,28], the normalized hypergraph Laplacian is:

$$\mathbf{L} = \mathbf{I}_N - \mathbf{D}_v^{-\frac{1}{2}} \mathbf{HMD}_e^{-1} \mathbf{H}^T \mathbf{D}_v^{-\frac{1}{2}}, \tag{3}$$

where $\mathbf{I}_N$ is an identity matrix.

2.2 The GCN Model

GCN has one input layer, two hidden layers and one output layer in [15]. The forward propagation of a two-layer GCN is:

$$\mathbf{Z} = \text{softmax}(\hat{\mathbf{A}} \text{ReLU}(\hat{\mathbf{A}} \mathbf{X} \mathbf{U}^{(0)}) \mathbf{U}^{(1)}), \tag{4}$$

where $\mathbf{Z} \in \mathbb{R}^{n \times c}$ is the output of the prediction label and $\mathbf{X} \in \mathbb{R}^{n \times d}$, $\hat{\mathbf{A}} = \tilde{\mathbf{D}}^{-\frac{1}{2}} \tilde{\mathbf{A}} \tilde{\mathbf{D}}^{\frac{1}{2}}$, $\mathbf{D}$ is a diagonal matrix with $\tilde{\mathbf{d}}_i = \sum_{j=1}^{n} \tilde{\mathbf{A}}_{ij}$ and $\tilde{\mathbf{A}} = \mathbf{A} + \mathbf{I}_N$ is an adjacency matrix. $\mathbf{U}^{(0)}$ and $\mathbf{U}^{(1)}$ are the weight matrices. And ReLU and softmax are the different activation functions. The final cross entropy loss function is:

$$\text{Loss} = -\sum_{i \in \mathbf{O}} \sum_{j=1}^{c} y_{ij} \ln z_{ij}, \tag{5}$$

where $\mathbf{O}$ is a set of labeled nodes.

3 Methodology

3.1 Our Proposed Method

In this paper, a more suitable graph matrix is learned by retaining the neighborhood structure of data. Given the data set $\mathbf{X} = \{\mathbf{x}_1, \mathbf{x}_2, ..., \mathbf{x}_n\} \in \mathbb{R}^{n \times d}$, where the i-th sample $\mathbf{x}_i$ and all data points can be connected to $\mathbf{x}_i$ as a neighbor with probability s_{ij}. When the distance $\|\mathbf{x}_i - \mathbf{x}_j\|_2^2$ is smaller, probability s_{ij} is larger. Besides, there are a lot of noise in the collected real data, which will affect the graph matrix [7,27]. Therefore, we introduce a sparse model to reduce the impact of noise points on the model [32]. We have:

$$\min_{\mathbf{S}, \mathbf{W}} \sum_{i,j} \|\mathbf{x}_i \mathbf{W} - \mathbf{x}_j \mathbf{W}\|_2^2 s_{ij} + \alpha \|\mathbf{W}\|_{2,1}$$
$$s.t. \quad \mathbf{s}_i^T \mathbf{1} = 1, s_{ij} \geq 0, \mathbf{W}^T \mathbf{W} = \mathbf{I}, \tag{6}$$

where α is a non-negative tuning parameter, and $\mathbf{W} \in \mathbb{R}^{d \times c}$ is the projection matrix. We add the constraint $\mathbf{s}_i^T \mathbf{1} = 1$ because some row vectors of $\mathbf{S}$ can not be all zero. We also use the $\ell_{2,1}$-norm for sparse learning to eliminate redundant information and irrelevant attributes [3,19].

Given the initial graph $\mathbf{A} \in \mathbb{R}^{n \times n}$, which is constructed by hypergraph. We learn a graph $\mathbf{S}$, which has a connected component c, and c is the number of classes [17]. We also use a low-rank constraint $rank(\mathbf{L_S}) = n - c$, which the corresponding similarity matrix has block diagonal [18,30]. We use F-norm to estimate the remainder:

$$\min_{\mathbf{S}} \|\mathbf{S} - \mathbf{A}\|_F^2$$
$$s.t. \quad rank(\mathbf{L_S}) = n - c, \mathbf{s}_i^T \mathbf{1} = 1, s_{ij} \geq 0, \tag{7}$$

where $\mathbf{L_S}$ is the Laplacian matrix.

We combine the above two goals and integrate them into a unified framework due to the influence of noise and outliers in the data. The final objective function is obtained:

$$\min_{\mathbf{S},\mathbf{W}} \sum_{i,j} \|\mathbf{x}_i \mathbf{W} - \mathbf{x}_j \mathbf{W}\|_2^2 s_{ij} + \alpha \|\mathbf{W}\|_{2,1} + \beta \|\mathbf{S} - \mathbf{A}\|_F^2$$
$$s.t. \quad rank(\mathbf{L_S}) = n - c, \mathbf{s}_i^T \mathbf{1} = 1, s_{ij} \geq 0, \mathbf{W}^T \mathbf{W} = \mathbf{I}, \tag{8}$$

where $\mathbf{S}$ represents the learned suitable graph. The constraint $rank(\mathbf{L_S}) = n - c$ depends on $\mathbf{S}$, so noting that $\sigma_i(\mathbf{L_S})$ is defined as the i-th minimum eigenvalue of $\mathbf{L_S}$. When $\mathbf{L_S}$ is semidefinite, $\sigma_i(\mathbf{L_S}) \geq 0$. In line with Ky Fan's Theorem [5], we have:

$$\sum_{i=1}^{c} \sigma_i(\mathbf{L_S}) = \min_{\mathbf{F}^T \mathbf{F}=\mathbf{I}, \mathbf{F} \in \mathbb{R}^{n \times c}} tr(\mathbf{F}^T \mathbf{L_S} \mathbf{F}). \tag{9}$$

So Eq. (8) becomes:

$$\min_{\mathbf{S},\mathbf{W},\mathbf{F}} \sum_{i,j} \|\mathbf{x}_i \mathbf{W} - \mathbf{x}_j \mathbf{W}\|_2^2 s_{ij} + \alpha \|\mathbf{W}\|_{2,1} + \beta \|\mathbf{S} - \mathbf{A}\|_F^2 + 2\tau tr(\mathbf{F}^T \mathbf{L_S} \mathbf{F})$$
$$s.t. \quad \mathbf{s}_i^T \mathbf{1} = 1, s_{ij} \geq 0, \mathbf{F}^T \mathbf{F} = \mathbf{I}, \mathbf{F} \in \mathbb{R}^{n \times c}, \mathbf{W}^T \mathbf{W} = \mathbf{I}, \tag{10}$$

where β and τ are non-negative tuning parameters, and $\mathbf{W}^T \mathbf{W} = \mathbf{I}$ represents that the feature space is distinctive after reduction [17,18]. By solving Eq. (11), we get a more suitable $\mathbf{S}$ for the GCN classification.

4 Experiment

4.1 Datasets

For verifying the robustness of our method, we use datasets of different types in Table 1. Datasets are obtained from UCI database[1].

---

[1] http://archive.ics.uci.edu/ml/.

Table 1. Descriptions of different datasets.

Datasets	Samples	Features	Classes
Isolet	1560	617	2
MnistData05	3495	784	10
MinistData10	6996	784	10
Pixraw10P	100	10000	10
Wall-following	5456	24	4
Zoo	101	16	7

4.2   Experimental Setting

For the convenience of subsequent experiments, we adopt a unified division method for all datasets, that is, in the training process, we adopt 10%, 20% and 30% of the labeled samples to construct and evaluate three different models. Moreover, we select 10% of the samples as the verification. Correspondingly the remaining 80%, 70% and 60% samples are used for testing. We use different data splits to average the results reported for 10 times.

Similar to the previous work [15]. We use a two-layer graph convolutional network and set the number of 16 hidden units. According to the Adam algorithm [14], the learning rate is set to 0.01. We use accuracy (ACC) as the evaluation index of experimental results, which is defined as:

$$\text{ACC} = \tfrac{N_r}{N}, \tag{11}$$

where N is the total number of samples, and N_r is the number of samples that can be correctly classified.

4.3   Comparison Methods

We compare the proposed method with two advanced methods, including hypergraph [26] and k-nearest neighbor (kNN) [20] graph respectively to compare with our graph method. And finally we add these constructed graphs into GCN, respectively.

- Hypergraph [26] includes a set of hyperedges, a set of vertices and a set of weights. In real life, data often contains multiple relationships, while the ordinary graph only contains binary relationships. So hypergraphs can learn more useful high-level correlations among data.
- k-nearest neighbor (kNN) [20] graph based on the similarity measurement among data. It calculates distance values of all pairs of nodes, and establishes new relationships between each node and its k nearest neighbors.

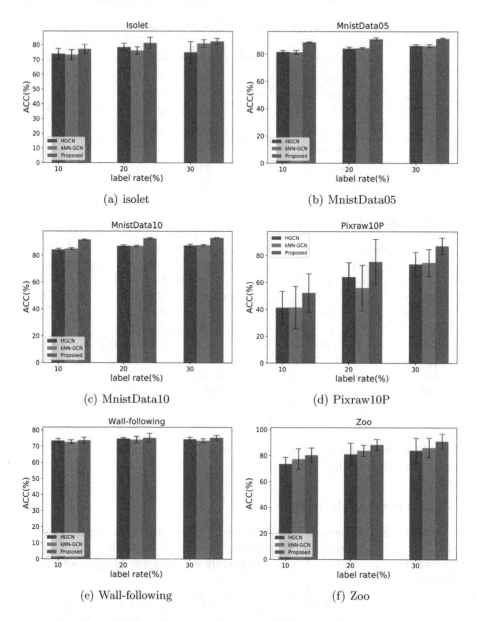

Fig. 1. Experimental results on different datasets.

4.4   Experimental Results

Figure 1 respectively lists the comparison results on datasets by three evaluation metrics. We can conclude that on high-dimensional datasets Mnistdata05 and Mnistdata10, when 10%, 20%, 30% samples are selected as training set, respectively, the proposed method has higher accuracy than the other two methods.

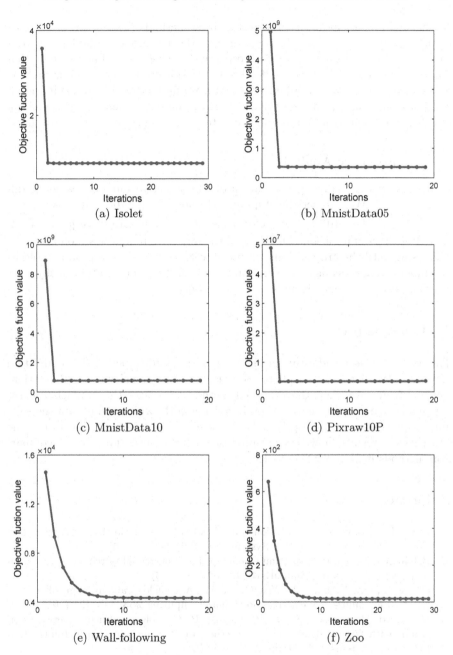

Fig. 2. The convergence curves of our objective function.

Especially on the dataset Pixraw10P, the proposed method has more than 11% improvement over HGCN when 10% samples are selected as the training set. On low dimensional datasets Zoo and Wall-following, our method also improves the accuracy compared with the other three methods. Especially, when 30% of samples are selected as the training set, the proposed method is superior to kNN-GCN by 4.9% and 1.92%, respectively. These experiments show that our proposed method is more effective in classification tasks. So our method is better because 1) a new adaptive graph learning method is more suitable for semi-supervised classification of GCNs; 2) the proposed graph method is superior to other graph construction methods due to the influence of noise and outliers.

4.5   Convergence Analysis

Figure 2 shows the convergence curves of the objective function value on different datases. We also set the stopping criterion of our proposed algorithm as $\frac{\|obj(t+1)-obj(t)\|_2^2}{obj(t)} < 10^{-5}$, where $obj(t)$ means the value of our objective function after the t-th iteration. The value of the objective function is monotonically decreasing until the proposed algorithm converges. Besides, the proposed objective function conversions in 30 iterations. So the fast convergence of the objective function proves the effectiveness of our method.

5   Conclusions

This paper proposes adaptive graph learning for semi-supervised classification of GCNs. We use hypergraph to establish the initial neighborhood relationship between data. And the proposed method integrates the Laplacian learning, sparse learning and hypergraph in a framework, which can obtain the suitable graph that more appropriates GCN for semi-supervised classifications. The experimental results show that our proposed method outperforms other comparison methods in classification tasks.

References

1. Atwood, J., Towsley, D.: Diffusion-convolutional neural networks. In: Advances in Neural Information Processing Systems, pp. 1993–2001 (2016)
2. Belkin, M., Niyogi, P.: Laplacian eigenmaps for dimensionality reduction and data representation. Neural Comput. **15**, 1373–1396 (2003)
3. Cheng, X., Zhu, Y., Song, J., Wen, G., He, W.: A novel low-rank hypergraph feature selection for multi-view classification. Neurocomputing **253**, 115–121 (2017)
4. Defferrard, M., Bresson, X., Vandergheynst, P.: Convolutional neural networks on graphs with fast localized spectral filtering. In: Advances in Neural Information Processing Systems, vol. 29, pp. 3844–3852 (2016)
5. Fan, K.: On a theorem of Weyl concerning eigenvalues of linear transformations I. Proc. Nat. Acad. Sci. U.S.A. **35**(11), 652 (1949)
6. Fu, S., Liu, W., Zhou, Y., Nie, L.: HpLapGCN: hypergraph p-Laplacian graph convolutional networks. Neurocomputing **362**, 166–174 (2019)

7. Gao, X., Hu, W., Guo, Z.: Exploring structure-adaptive graph learning for robust semi-supervised classification. In: 2020 IEEE International Conference on Multimedia and Expo, pp. 1–6. IEEE (2020)
8. Grover, A., Leskovec, J.: node2vec: scalable feature learning for networks. In: Proceedings of the 22nd ACM SIGKDD International Conference on Knowledge Discovery and Data Mining, pp. 855–864 (2016)
9. Guo, Y., Wu, Z., Shen, D.: Learning longitudinal classification-regression model for infant hippocampus segmentation. Neurocomputing **391**, 191–198 (2020)
10. Hao, S., Zhou, Y., Guo, Y.: A brief survey on semantic segmentation with deep learning. Neurocomputing **406**, 302–321 (2020)
11. Hu, R., Zhu, X., Zhu, Y., Gan, J.: Robust SVM with adaptive graph learning. World Wide Web **23**(3), 1945–1968 (2020)
12. Jiang, B., Zhang, Z., Lin, D., Tang, J., Luo, B.: Semi-supervised learning with graph learning-convolutional networks. In: Proceedings of the IEEE Conference on Computer Vision and Pattern Recognition, pp. 11313–11320 (2019)
13. Kang, Z., Pan, H., Hoi, S.C., Xu, Z.: Robust graph learning from noisy data. IEEE Trans. Cybern. **50**(5), 1833–1843 (2019)
14. Kingma, D.P., Ba, J.: Adam: a method for stochastic optimization. In: International Conference on Learning Representations (2015)
15. Kipf, T., Welling, M.: Semi-supervised classification with graph convolutional networks. In: International Conference on Learning Representations (2017)
16. Li, Y., Zhang, S., Cheng, D., He, W., Wen, G., Xie, Q.: Spectral clustering based on hypergraph and self-re-presentation. Multimed. Tools Appl. **76**(16), 17559–17576 (2016). https://doi.org/10.1007/s11042-016-4131-6
17. Nie, F., Wang, X., Jordan, M.I., Huang, H.: The constrained Laplacian rank algorithm for graph-based clustering. In: Thirtieth AAAI Conference on Artificial Intelligence (2016)
18. Nie, F., Wei, Z., Li, X.: Unsupervised feature selection with structured graph optimization. In: Thirtieth AAAI Conference on Artificial Intelligence (2016)
19. Niepert, M., Ahmed, M., Kutzkov, K.: Learning convolutional neural networks for graphs. In: International Conference on Machine Learning, pp. 2014–2023 (2016)
20. Peterson, L.E.: K-nearest neighbor. Scholarpedia **4**(2), 1883 (2009)
21. Qian, X., Huang, H., Chen, X., Huang, T.: Efficient construction of sparse radial basis function neural networks using L1-regularization. Neural Netw. **94**, 239–254 (2017)
22. Shen, H.T., et al.: Heterogeneous data fusion for predicting mild cognitive impairment conversion. Inf. Fusion **66**, 54–63 (2021). https://doi.org/10.1016/j.inffus.2020.08.023
23. Shen, H.T., Zhu, Y., Zheng, W., Zhu, X.: Half-quadratic minimization for unsupervised feature selection on incomplete data. IEEE Trans. Neural Netw. Learn. Syst. (2020). https://doi.org/10.1109/TNNLS.2020.3009632
24. Veličković, P., Cucurull, G., Casanova, A., Romero, A., Liò, P., Bengio, Y.: Graph attention networks. In: International Conference on Learning Representations (2018). https://openreview.net/forum?id=rJXMpikCZ
25. Wang, D., Cui, P., Zhu, W.: Structural deep network embedding. In: Proceedings of the 22nd ACM SIGKDD International Conference on Knowledge Discovery and Data Mining, pp. 1225–1234 (2016)
26. Yadati, N., Nimishakavi, M., Yadav, P., Louis, A., Talukdar, P.: HyperGCN: hypergraph convolutional networks for semi-supervised classification. arXiv preprint arXiv:1809.02589 (2018)

27. Zhao, Z., Liu, H.: Spectral feature selection for supervised and unsupervised learning. In: Proceedings of the 24th International Conference on Machine Learning, pp. 1151–1157 (2007)
28. Zhou, D., Huang, J., Schölkopf, B.: Learning with hypergraphs: clustering, classification, and embedding. In: Advances in Neural Information Processing Systems, pp. 1601–1608 (2007)
29. Zhou, Y., Tian, L., Zhu, C., Jin, X., Sun, Y.: Video coding optimization for virtual reality 360-degree source. IEEE J. Sel. Top. Signal Process. **14**(1), 118–129 (2019)
30. Zhu, X., Gan, J., Lu, G., Li, J., Zhang, S.: Spectral clustering via half-quadratic optimization. World Wide Web **23**(3), 1969–1988 (2019). https://doi.org/10.1007/s11280-019-00731-8
31. Zhu, X., et al.: Joint prediction and time estimation of Covid-19 developing severe symptoms using chest CT scan. Med. Image Anal. **67**, 101824 (2021)
32. Zhu, X., Zhang, S., Zhu, Y., Zhu, P., Gao, Y.: Unsupervised spectral feature selection with dynamic hyper-graph learning. IEEE Trans. Knowl. Data Eng. (2020). https://doi.org/10.1109/TKDE.2020.3017250

Semi-supervised Feature Selection Based on Cost-Sensitive and Structural Information

Yiling Tao, Guangquan Lu$^{(\boxtimes)}$, Chaoqun Ma, Zidong Su, and Zehui Hu

Guangxi Key Lab of Multi-Source Information Mining and Security,
Guangxi Normal University, Guilin, Guangxi, China
`lugq@mailbox.gxnu.edu.cn`

Abstract. Feature selection is an important process of high-dimensional data analysis in data mining and machine learning. In the feature selection stage, the cost of misclassification and the structural information of paired samples on each feature dimension are often ignored. To overcome this, we propose semi-supervised feature selection based on cost-sensitive and structural information. First, cost-sensitive learning is incorporated into the semi-supervised framework. Second, the structural information between a pair of samples in each feature dimension is encapsulated into the feature graph. Finally, the correlation between the candidate feature and the target feature is added, which avoids the misunderstanding of the feature with low correlation as the salient feature. Furthermore, the proposed method also considers the redundancy between feature pairs, which can improve the accuracy of feature selection. The proposed method is more interpretable and practical than previous semi-supervised feature selection algorithms, because it considers the misclassification cost, structural relationship and the correlations between features and target features. Experimental results show that the promising performance of the proposed method outperforms the state-of-the-arts on eight data sets.

Keywords: Feature selection · Cost-sensitive · Structural relationship · Semi-supervised

1 Introduction

Big data has widely appeared in various fields, such as pattern recognition and machine learning [1,2]. A common problem in data processing is that the data often contains some unimportant features [3,4], which will increase the calculation cost and affect the effectiveness of model training [5,6]. Therefore, feature selection has become one of the important research fields of machine learning in recent years.

Feature selection is used to delete redundant features for conducting dimensionality reduction [1], which can help model training and reduce the impact of "dimension disaster" [7]. Depending on the availability of sample labels, feature

© Springer Nature Switzerland AG 2021
M. Qiao et al. (Eds.): ADC 2021, LNCS 12610, pp. 23–36, 2021.
https://doi.org/10.1007/978-3-030-69377-0_3

selection is divided into supervised, semi-supervised and unsupervised. Supervised feature selection [8] only uses labeled samples to train the model, and takes the structural relationship between labels and features to choose the important features, so as to explores the result of feature subset with the highest relevance to the label. Unsupervised feature selection [9–11] uses unlabeled sample training model, which selects the most representative features from the original feature set according to certain evaluation criteria. Semi-supervised feature selection [12,13] uses a small number of labeled samples and a lot of unlabeled samples to achieve the optimal feature subset. These types of methods are efficient, because they can not only mine the global and local structure of all samples, but also utilize the small number of labels that providing category information. Therefore, this paper focus on research on semi-supervised feature selection.

Various semi-supervised feature selection methods have been proposed recently. For example, the typical feature selection based on classifier [14] (semi-supervised support vector machine, S3VM), it uses support vector machine (SVM) to tag no label samples, and then fused the "soft" label samples for model training. Zhao and Liu [15] proposed a semi-supervised regularized feature selection framework based on spectral learning to evaluate the correlation of features. In addition, Ren et al. [16] proposed a forward semi-supervised feature selection framework based on wrapper type, which combines forward selection with wrapper type to obtain the optimal feature subset. Chen et al. [17] combined the traditional fisher-score method to obtain the global optimal feature subset by the "soft" label of unlabeled samples with label propagation technology.

However, the existing semi-supervised feature selection methods have some defects. First, many semi-supervised feature selection researches focus on the lowest classification error rate without considering the misclassification cost. It has assumed that different misclassifications owning the equal costs [18], which may lead the model pays attention to samples which causes high misclassification losses, resulting in biases in the features selected by the learning model. Second, some advanced semi-supervised feature selection algorithms do not consider the structural information of the paired samples in each feature dimension, which can improve the performance of feature selection [19]. In addition, researchers believe that the correlation of a single candidate features is equal to the correlation of selected features, without considering the joint correlation of a pair of features, which will regard low-relevance features as salient features. Therefore, some low-correlation features are regarded as salient features.

To solve the above problems, we propose semi-supervised feature selection based on cost-sensitive and structural information (SF_CSSI). The contributions of this paper are as follows:

- In practical applications, misclassification has always existed, however, the cost of misclassification is always ignored by researchers. The proposed method considers the misclassification cost and sets different penalty costs for different categories samples. In contrast to conventional feature selection methods, we try to minimize the total cost rather than the total error rate, aiming to prevent disasters caused by mistakes with high costs.

- In this paper, the proposed method converts each original feature vector into a structure-based feature graph representation, which contains structural information between sample pairs in each feature dimension, in order to preserve more meaningful information. Furthermore, the proposed method constructs feature information matrix to simultaneously maximize joint relevancy of different pairwise feature combinations in relation to the target feature graphs and minimize redundancy among selected features, so as to obtain feature subset with high correlation and low redundancy.
- The method proposed in this paper has rarely been studied, because it considers misclassification cost, structural information and information measurement of paired features. Experiments prove that the proposed method in this paper can achieve better feature selection results on real datasets.

2   Approach

2.1   Notations

In this paper, matrices are written as boldface uppercase letters, vectors are written as boldface lowercase letters and scalars are written as normal italic letters. For matrix $\mathbf{X}$, $x_{i,j}$ represents the element in the i-th row and j-th column of $\mathbf{X}$. The Frobenius norm of matrix $\mathbf{X} \in \mathbb{R}^{n \times d}$ is defined as $\|\mathbf{X}\|_F = \sqrt{\sum_{i,j} x_{i,j}^2}$. The $l_{2,1}$-norm of matrix $\mathbf{X}$ is defined as $\|\mathbf{X}\|_{2,1} = \sum_{i=1}^{n} \sqrt{\sum_{j=1}^{d} x_{ij}^2}$. For vector $\mathbf{x}$, its l_1-norm is defined $\|\mathbf{x}\|_1 = \sum_{i=1}^{n} |x_i|$. The symbol $\odot$ denotes multiplication of corresponding elements and $tr(\mathbf{X})$ represents the trace of matrix $\mathbf{X}$.

In semi-supervised learning, the data set consists of two parts: labeled data $\mathbf{X}_L = (x_1, x_2, \dots, x_l)$ and unlabeled data $\mathbf{X}_U = (x_{l+1}, x_{l+2}, \dots, x_{l+u})$, $u = n - l$, n represents the number of samples, l represents the number of labeled samples, u represents the number of unlabeled samples. The corresponding labels is $\mathbf{Y}_L = (y_1, y_2, \dots, y_l)^T$ and the label of $\mathbf{Y}_U = (y_{l+1}, y_{l+2}, \dots, y_{l+u})^T$ is unknown.

2.2   Cost-Sensitive Feature Selection

Given data set $\mathbf{X} = [\mathbf{x}_1, \mathbf{x}_2, \cdots, \mathbf{x}_n] \in \mathbb{R}^{n \times d}$, n represents the number of samples, d represents the features of each sample. The traditional feature selection imposes a sparsity penalty in the objective function, which makes the selected features more sparse and more discriminative. The objective function of traditional feature selection [9] is defined as:

$$\min_{\mathbf{W}} \|\mathbf{Y} - \mathbf{XW}\|_F^2 + \lambda \|\mathbf{W}\|_{2,1} \tag{1}$$

However, cost-sensitive learning is embedded into feature selection framework because the misclassification problem often occurs in practical applications. Cost-sensitive learning assigns different cost parameters to different types of samples, without loss of generality, so the specified cost matrix is introduced into the feature selection framework. The traditional cost-sensitive feature selection objective function [20] is defined as:

$$\min_{\mathbf{W}} \left\| (\mathbf{X}^T \mathbf{W} - \mathbf{Y}) \odot \mathbf{C} \right\|_{2,1} + \lambda \|\mathbf{W}\|_{2,1}, \tag{2}$$

where $\mathbf{W} \in \mathbb{R}^{d \times m}$ represents the feature weight matrix, $\mathbf{Y} \in \mathbb{R}^{n \times m}$ represents labels, $\mathbf{C} \in \mathbb{R}^{n \times m}$ represents cost matrix, λ represents the penalty coefficient.

2.3   Feature Selection with Graph Structural Information

The structural information can provide more abundant representation but few researchers pay attention to these between the features in each pairs of samples.

Therefore, each feature vector is transformed into a feature graph structure, which encapsulates the pairwise relationship between samples. In addition, the information theory criterion of Jensen-Shannon divergence is used to measure the joint correlation between different paired feature combinations and target labels. The specific process is as follows.

Let $\mathbf{X} = \{\mathbf{f}_1, \ldots, \mathbf{f}_i, \ldots, \mathbf{f}_N\} \in \mathbb{R}^{M \times N}$ represents a data set of M samples and N features. Each original feature vector $\mathbf{f}_i = (f_{i1}, \ldots, f_{ia}, \ldots, f_{ib}, \ldots, f_{iM})^T$ is transformed into a feature graph $\mathbf{G}_i(V_i, E_i)$, where vertex $v_{ia} \in V_i$ represents the a-th sample f_{ia} in feature $\mathbf{f}_i$ (i.e., each vertex represents a sample), edge $(v_{ia}, v_{ib}) \in E_i$ represents the weight of the a-th sample and the b-th sample (i.e., the edge represents the correlation between a pair of samples in the corresponding feature dimension). In addition, we also construct a graph structure for the target feature $\mathbf{Y}$. For classification problems, $\mathbf{Y}$ are discrete value $c \in \{1, 2, \ldots, m\}$. Therefore, we calculate the continuous value of each discrete target feature $\mathbf{f}_i$ as $\hat{\mathbf{f}}_i = \left(\hat{f}_{i1}, \ldots, \hat{f}_{ia}, \ldots, \hat{f}_{ib}, \ldots, \hat{f}_{iM} \right)^T$, $\hat{f}_{ia}$ represents the a-th sample in $\hat{\mathbf{f}}_i$. When the f_{ia} in $\mathbf{f}_i$ belongs to class m, $\hat{f}_{ia}$ is the mean value of all class m samples in $\mathbf{f}_i$. Similarly, we construct the graph structure of the target feature $\hat{\mathbf{f}}_i$ as $\hat{\mathbf{G}}_i \left(\hat{V}_i, \hat{E}_i \right)$. $\hat{v}_{ia} \in \hat{V}_i$ represents the a-th sample in target feature $\hat{\mathbf{f}}_i$, $\left(\hat{v}_{ia}, \hat{v}_{ib} \right) \in \hat{E}_i$ is the weighted edge connecting the a-th sample and the b-th sample of $\hat{\mathbf{f}}_i$. This paper uses Euclidean distance to calculate the relationship between pairs of feature samples, that is, the weight of f_{ia} and f_{ib} can be expressed as:

$$\omega(v_{ia}, v_{ib}) = \sqrt{(f_{ia} - f_{ib})^2} \tag{3}$$

Similarly, the weight of edge $\left(\hat{v}_{ia}, \hat{v}_{ib} \right) \in \hat{E}_i$ in $\hat{\mathbf{G}}_i \left(\hat{V}_i, \hat{E}_i \right)$ is expressed as follows:

$$\omega\left(\hat{v}_{ia}, \hat{v}_{ib} \right) = \sqrt{(\mu_{ia} - \mu_{ib})^2}, \tag{4}$$

where μ_{ia} is the mean value of all samples in $\mathbf{f}_i$ from the same class m.

Jensen Shannon divergence (JSD) is used to measure the divergence between two probability distributions [21]. Give two (discrete) probability distributions $\mathcal{P} = (p_1, \ldots, p_a, \ldots p_A)$ and $\mathcal{K} = (k_1, \ldots, k_b, \ldots k_B)$. The JSD between $\mathcal{P}$ and $\mathcal{K}$ is defined as:

$$D_{\mathrm{JS}}\left(\mathcal{P}, \mathcal{K}\right) = H_S\left(\frac{\mathcal{P}+\mathcal{K}}{2}\right) - \frac{1}{2}H_S\left(\mathcal{P}\right) - \frac{1}{2}H_S\left(\mathcal{K}\right), \tag{5}$$

where $H_S\left(\mathcal{P}\right) = \sum_{i=1}^{A} p_i \log p_i$ is the Shannon entropy of probability distribution $\mathcal{P}$. In the literature [22], the JSD has been used as a means of measuring the information theoretic dissimilarity between graphs associated with their probability distributions. In this paper, we focus on the similarity between graph-based feature representations. We use the negative exponent of $D_{\mathrm{JS}}\left(\mathcal{P}, \mathcal{K}\right)$ to calculate the similarity I_S between probability distributions $\mathcal{P}$ and $\mathcal{K}$, so:

$$I_S\left(\mathcal{P}, \mathcal{K}\right) = \exp\left\{-D_{\mathrm{JS}}\left(\mathcal{P}, \mathcal{K}\right)\right\} \tag{6}$$

The information theoretic function is used to evaluate the relevance between different feature combination and target labels to achieve the maximum correlation and minimum redundancy standards. For a set of N features $\mathbf{f}_1, \ldots, \mathbf{f}_i, \ldots, \mathbf{f}_N$ and related continuous target feature $\mathbf{Y}$, the correlation degree of feature pair $\{\mathbf{f}_i, \mathbf{f}_j\}$ is expressed as follows:

$$U_{f_i, f_j} = \frac{I_s\left(\mathbf{G}_i, \overset{\wedge}{\mathbf{G}}\right) + I_s\left(\mathbf{G}_j, \overset{\wedge}{\mathbf{G}}\right)}{I_s\left(\mathbf{G}_i, \mathbf{G}_j\right)}, \tag{7}$$

where I_s is the JSD based similarity measure of information theory defined in Eq. 6. $I_s\left(\mathbf{G}_i, \overset{\wedge}{\mathbf{G}}\right)$ represents the correlation measures of feature $\mathbf{f}_i$ with target feature $\mathbf{Y}$. $I_s\left(\mathbf{G}_j, \overset{\wedge}{\mathbf{G}}\right)$ represents the correlation measures of feature $\mathbf{f}_j$ with target feature $\mathbf{Y}$. $I_s\left(\mathbf{G}_i, \mathbf{G}_j\right)$ denotes the redundancy of paired feature $\{\mathbf{f}_i, \mathbf{f}_j\}$. Therefore, U_{f_i, f_j} is large if and only if $I_s\left(\mathbf{G}_i, \overset{\wedge}{\mathbf{G}}\right) + I_s\left(\mathbf{G}_j, \overset{\wedge}{\mathbf{G}}\right)$ is large and $I_s\left(\mathbf{G}_i, \mathbf{G}_j\right)$ is small. This indicates that the pairwise feature $\{\mathbf{f}_i, \mathbf{f}_j\}$ is informative and less redundant.

Given the feature information matrix $\mathbf{U}$ and d-dimensional feature vector $\mathbf{w}$. The feature subset is identified by solving the maximization problem of the following formula:

$$\max f(\mathbf{w}) = \max_{\mathbf{w} \in \mathbb{R}^d} \mathbf{w}^T \mathbf{U} \mathbf{w}, \tag{8}$$

where $\mathbf{w} \in \mathbb{R}^d$, $\mathbf{w} = (w_1, w_2, \cdots, w_i, \cdots, w_n)^T$, $w_i > 0$, w_i represents the correlation coefficient of the i-th feature.

2.4   Mathematical Formulation

The purpose of our proposed method is to improve the performance of feature selection through structural information and misclassification costs when the data does not have a large number of labels. Therefore, we combine cost-sensitive and Eq. 8 to propose semi-supervised feature selection based on cost-sensitive and structural information. The specific mathematical expression is as follows:

$$\min_{\mathbf{w}} \alpha_1 tr(\mathbf{w}^T \mathbf{X}^T \mathbf{L} \mathbf{X} \mathbf{w}) + \sum_{i=1}^{l} \|x_i \mathbf{w} - y_i\|_2^2 c_i + \alpha_2 \|\mathbf{w}\|_1 - \alpha_3 \mathbf{w}^T \mathbf{U} \mathbf{w} \quad (9)$$

The first term represents the learning of local proximity structure, which helps the model to select a representative feature subset by maintaining the local structure of the samples. $\mathbf{w}$ represents the feature coefficient vector, $\mathbf{L}$ is the Laplacian matrix, $\mathbf{L} = \mathbf{D} - \mathbf{A}$, where $\mathbf{D}$ is a diagonal matrix, the diagonal element satisfies $D_{ii} = \sum_{j=1}^{n} A_{ij}$ and $\mathbf{A}$ is the affinity matrix, if $i \neq j$, $A_{ij} = \exp(-\frac{\|x_i - x_j\|_2^2}{2\sigma^2})$; otherwise, $A_{ij} = 0$. The cost c_i represents the cost, the second term indicates that the loss of the original feature is combined with the cost to obtain the misclassification cost loss. Since we judge the misclassification result based on the label sample, we only use the labeled sample to calculate the misclassification loss. The third term $\|\mathbf{w}\|_1$ represents the sparse regular term, which uses the l_1-norm to shrink some coefficients to zero. The fourth term encourages the selected features to be jointly more relevant with the target while maintaining less redundancy between features, α_1, α_2 and α_3 are the penalty coefficients.

2.5   Optimization

In order to optimize, Eq. 9 can be rewritten as follows:

$$\min_{\mathbf{w}, \mathbf{Q}} \alpha_1 tr\left(\mathbf{w}^T \mathbf{X}^T \mathbf{L} \mathbf{X} \mathbf{w}\right) + tr\left((\mathbf{X}_L \mathbf{w} - \mathbf{Y}_L)^T \mathbf{C} (\mathbf{X}_L \mathbf{w} - \mathbf{Y}_L)\right)$$
$$+ \alpha_2 tr\left(\mathbf{w}^T \mathbf{Q} \mathbf{w}\right) - \alpha_3 \mathbf{w}^T \mathbf{U} \mathbf{w}, \quad (10)$$

where $\mathbf{Q}$ is the diagonal matrix. We use the idea of iterative learning to optimize the objective function, that is, update $\mathbf{w}$ by fixing $\mathbf{Q}$ and update $\mathbf{Q}$ by fixing $\mathbf{w}$, until Eq. 9 converges, so that the optimal solution of weight vector $\mathbf{w}$ can be obtained.

– Update $\mathbf{w}$ by fixing $\mathbf{Q}$

When $\mathbf{Q}$ is fixed, Eq. 10 can be regarded as a function of $\mathbf{w}$:

$$L(\mathbf{w}) = \alpha_1 tr\left(\mathbf{w}^T \mathbf{X}^T \mathbf{L} \mathbf{X} \mathbf{w}\right) + tr\left((\mathbf{X}_L \mathbf{w} - \mathbf{Y}_L)^T \mathbf{C} (\mathbf{X}_L \mathbf{w} - \mathbf{Y}_L)\right)$$
$$+ \alpha_2 tr\left(\mathbf{w}^T \mathbf{Q} \mathbf{w}\right) - \alpha_3 \mathbf{w}^T \mathbf{U} \mathbf{w} \quad (11)$$

We take the derivative of $\mathbf{w}$ in Eq. 11 and make it equal to zero:

$$\frac{\partial L}{\partial \mathbf{w}} = 2\alpha_1 \mathbf{X}^T \mathbf{L} \mathbf{X} \mathbf{w} + 2\mathbf{X}_L{}^T \mathbf{C} \mathbf{X}_L \mathbf{w} - 2\mathbf{X}_L{}^T \mathbf{C} \mathbf{Y}_L + 2\alpha_2 \mathbf{Q} \mathbf{w} - 2\alpha_3 \mathbf{U} \mathbf{w} = 0 \quad (12)$$

According to Eq. 12, it is solved as follows:

$$\mathbf{w} = \left(2\alpha_1 \mathbf{X}^T \mathbf{L} \mathbf{X} + 2\mathbf{X}_L{}^T \mathbf{C} \mathbf{X}_L + 2\alpha_2 \mathbf{Q} - 2\alpha_3 \mathbf{U}\right)^{-1} 2\mathbf{X}_L{}^T \mathbf{C} \mathbf{Y}_L \qquad (13)$$

- Update $\mathbf{Q}$ by fixing $\mathbf{w}$

When $\mathbf{w}$ is fixed, Eq. 10 can be regarded as:

$$\min_{\mathbf{Q}} \alpha_2 tr\left(\mathbf{w}^T \mathbf{Q} \mathbf{w}\right) \qquad (14)$$

By setting the partial derivative of the above function with respect to $\mathbf{Q}$ as 0 and according to the article [23], it is solved as follows:

$$Q_{ii} = \frac{1}{2\,|w_i|}, \qquad (15)$$

where $\mathbf{Q}$ is a diagonal matrix and $Q_{ii} = \frac{1}{2|w_i|}$ is the diagonal element.

Algorithm 1: The pseudo code of solving Eq. 9

Input: Data matrix $\mathbf{X} \in \mathbb{R}^{n \times d}$, labeled data $\mathbf{X}_L \in \mathbb{R}^{l \times d}$, labels $\mathbf{Y}_L \in \mathbb{R}^l$, cost matrix $\mathbf{C} \in \mathbb{R}^{l \times l}$,
        control parameters $\alpha_1, \alpha_2, \alpha_3$;
Output: $\mathbf{w} \in \mathbb{R}^d$;
1. Initialize $t = 0$ and $\mathbf{Q}^{(0)}$;
2. Build affinity matrix $\mathbf{A}$, $A_{ij} = \exp(-\frac{||x_i - x_j||_2^2}{2\sigma^2})$ or $A_{ij} = 0$;
3. Build diagonal matrix $\mathbf{D}$, $D_{ii} = \sum_{j=1}^{n} A_{ij}$;
4. Build Laplacian matrix $\mathbf{L}$, $\mathbf{L} = \mathbf{D} - \mathbf{A}$;
5. **repeat:**
   5.1 Update $\mathbf{w}^{(t+1)}$ via Eq. 13;
   5.2 Update $\mathbf{Q}^{(t+1)}$ via Eq. 15;
   5.3 $t = t + 1$;
until converges;

2.6   Convergence Analysis

Let $\mathbf{w}$ and $\mathbf{Q}$ be $\mathbf{w}^{(t)}$ and $\mathbf{Q}^{(t)}$ in the t-th iteration, and Eq. 10 can be rewritten as:

$$E\left(\mathbf{w}^{(t)}, \mathbf{Q}^{(t)}\right) = \alpha_1 tr\left(\left(\mathbf{w}^{(t)}\right)^T \mathbf{X}^T \mathbf{L} \mathbf{X} \mathbf{w}^{(t)}\right) + tr\left(\left(\mathbf{X}_L \mathbf{w}^{(t)} - \mathbf{Y}_L\right)^T \mathbf{C} \left(\mathbf{X}_L \mathbf{w}^{(t)} - \mathbf{Y}_L\right)\right)$$
$$+ \alpha_2 tr\left(\left(\mathbf{w}^{(t)}\right)^T \mathbf{Q}^{(t)} \mathbf{w}^{(t)}\right) - \alpha_3 \left(\mathbf{w}^{(t)}\right)^T \mathbf{U} \mathbf{w}^{(t)} \qquad (16)$$

Because the objective function $E\left(\mathbf{w}^{(t)}, \mathbf{Q}^{(t)}\right)$ is a convex optimization problem about $\mathbf{w}$, we have the following inequality:

$$E\left(\mathbf{w}^{(t+1)}, \mathbf{Q}^{(t)}\right) \leq E\left(\mathbf{w}^{(t)}, \mathbf{Q}^{(t)}\right) \qquad (17)$$

According to the article [23], we know that Eq. 14 is convergent, so we can deduce that Eq. 10 is convergent about $\mathbf{Q}$, so we express the convergence as the following inequality:

$$E\left(\mathbf{w}^{(t+1)},\mathbf{Q}^{(t+1)}\right) \leq E\left(\mathbf{w}^{(t+1)},\mathbf{Q}^{(t)}\right) \tag{18}$$

Combining Eq. 17 and Eq. 18, we can get the inequality:

$$E\left(\mathbf{w}^{(t+1)},\mathbf{Q}^{(t+1)}\right) \leq E\left(\mathbf{w}^{(t)},\mathbf{Q}^{(t)}\right) \tag{19}$$

Equation 16 is non-increasing at each iteration according to Eq. 19. Therefore, the proposed Algorithm 1 is convergence.

3   Experiments

In this section, we evaluated our proposed SF_CSSI and other six comparison methods on eight data sets. Specially, we first employed each feature selection method to choose the new feature subsets from original data sets, and then utilized support vector machine classification to evaluate the selected subsets.

3.1   Datasets and Comparison Methods

The data sets (i.e., madelon, SECOM, chess, isolet, Hill-with, Hill-without, musk and sonar) are from UCI Machine Learning Repository[1]. We summarized the detail of all data sets in Table 1.

Table 1. Summarization of data sets.

Datasets	Samples	Features	Classes
madelon	2000	500	2
SECOM	1967	590	2
chess	3196	36	2
isolet	1560	617	2
Hill-with	606	100	2
Hill-without	606	100	2
musk	486	166	2
sonar	208	360	2

We compared our proposed method with six comparison methods and the details of them are listed as follow:

– Cost-Sensitive Laplacian Score (CSLS [24]) uses Laplacian graphs and the cost of misclassification between classes to score each feature individually.

---

[1] http://archive.ics.uci.edu/ml/.

- Semi-supervised feature selection based on joint mutual information (Semi-JMI [25]) uses the redundancy between features and the correlation between features and labels to complete feature selection.
- Semi-supervised feature selection based on information theory method (Semi-IMIM [25]) only uses the correlation between features and labels to complete feature selection.
- Cost-Sensitive Feature Selection via F-Measure Optimization Reduction (CSFS [20]) introduces cost sensitivity to select features, which optimizes F-measure instead of accuracy to take class imbalance issue into account.
- Cost-sensitive feature selection via the $l_{2,1}$-norm (CSEFS [26]) combines $l_{2,1}$-norm minimization regularization and loss term of embedding misclassification cost to select feature subset.
- Semi-supervised Feature Selection via Rescaled Linear Regression (RLSR [17]) uses a set of scale factors to adjust regression coefficients, then uses regression coefficients to rank features.

3.2 Experimental Settings

The experiment of this paper is implemented with the MATLAB 2018a under Windows 10 system. Referring to [27] article's method, we can divide the data set into three parts: labeled sample set (L), unlabeled sample set (U), and test sample set (T). For each of data sets, the labeled samples were randomly selected with the given ratio $\{10\%, 20\%, 30\%\}$.

We use 10-fold cross-validation to generate training sample set and test sample set, then randomly select L and U from the training sample set for training, and finally use T to test the performance of different methods. All algorithms perform 10 times 10-fold cross-validation and take the average of the 10 experimental results as the final total cost, which reduce the accidental occurrence. We set the parameters α_1, α_2 and α_3 in Eq. 9 in range of $\{10^{-3}, 10^{-1}, ..., 10^1, 10^3\}$. For other comparison methods, we set these according to their corresponding literature.

Table 2. Total cost (cost ± std) of misclassification on eight data sets. Bold numbers indicate the best results.

Cost	Data sets	CSLS	Semi-JMI	Semi-IMIM	CSEFS	CSFS	RLSR	Proposed
$cost_1 = 10$ $cost_2 = 25$	madelon	1399.40 ± 17.57	1396.65 ± 20.88	1394.35 ± 19.39	1540.60 ± 74.59	1462.65 ± 19.22	1405.23 ± 23.38	**1369.56** ± 12.96
	SECOM	273.65 ± 5.87	270.26 ± 5.51	269.02 ± 5.19	268.03 ± 5.41	268.88 ± 5.70	269.01 ± 6.60	**266.09** ± 4.41
	chess	2494.00 ± 9.98	2349.15 ± 20.86	2624.30 ± 12.03	2485.75 ± 15.27	2726.00 ± 24.67	2074.65 ± 10.29	**515.00** ± 0.00
	isolet	525.50 ± 19.97	553.00 ± 22.84	538.90 ± 22.34	413.20 ± 19.60	352.05 ± 28.52	349.35 ± 15.91	**346.60** ± 10.53
	Hill-with	167.50 ± 18.73	177.38 ± 18.68	147.05 ± 63.65	157.35 ± 23.30	105.10 ± 17.08	102.75 ± 21.07	**100.40** ± 15.60
	Hill-without	1.05 ± 0.72	141.40 ± 52.04	168.30 ± 1.35	1.35 ± 1.22	2.50 ± 5.52	4.05 ± 5.90	**0.65** ± 0.89
	musk	165.20 ± 7.77	133.10 ± 7.94	145.75 ± 10.33	151.50 ± 11.77	132.30 ± 11.33	141.00 ± 17.88	**130.00** ± 6.10
	Sonar	99.20 ± 11.40	94.20 ± 6.17	93.95 ± 6.96	93.55 ± 7.26	99.00 ± 9.52	98.20 ± 8.79	**89.35** ± 4.77
$cost_1 = 25$ $cost_2 = 10$	madelon	1675.57 ± 14.98	1678.53 ± 19.80	1652.81 ± 12.71	1536.18 ± 25.85	1577.72 ± 23.29	1526.05 ± 18.59	**1510.15** ± 13.54
	SECOM	168.90 ± 27.01	161.10 ± 26.70	165.75 ± 30.30	163.45 ± 22.17	167.45 ± 21.50	**154.80** ± 19.11	158.40 ± 20.37
	chess	1193.80 ± 4.52	964.65 ± 8.23	1060.85 ± 7.13	1222.60 ± 34.73	1175.15 ± 7.43	**921.00** ± 7.32	1222.60 ± 34.73
	isolet	350.35 ± 11.60	432.90 ± 15.18	435.85 ± 12.80	334.40 ± 11.16	331.35 ± 8.08	321.85 ± 6.33	**318.76** ± 5.73
	Hill-with	533.10 ± 32.68	499.20 ± 54.94	464.15 ± 42.81	482.80 ± 53.24	482.80 ± 53.34	454.45 ± 68.49	**411.95** ± 43.52
	Hill-without	1115.20 ± 34.18	428.10 ± 55.74	427.26 ± 53.62	43.60 ± 32.47	2.35 ± 1.98	9.30 ± 1.99	**1.60** ± 2.50
	musk	150.15 ± 7.01	129.50 ± 9.66	133.95 ± 6.83	137.40 ± 9.81	128.95 ± 11.14	129.75 ± 12.04	**128.65** ± 11.09
	Sonar	168.90 ± 27.01	161.10 ± 26.70	165.75 ± 30.30	163.45 ± 22.17	167.45 ± 21.50	154.80 ± 19.11	**153.40** ± 20.37

Table 3. The value of specificity on eight data sets. Bold numbers indicate the best results.

Cost	Data sets	CSLS	Semi-JMI	Semi-IMIM	CSEFS	CSFS	RLSR	Proposed
$cost_1 = 10$ $cost_2 = 25$	madelon	61.34 ± 0.70	60.98 ± 0.86	60.41 ± 0.96	56.31 ± 0.18	58.19 ± 0.62	57.99 ± 0.38	**63.93 ± 1.96**
	SECOM	98.30 ± 0.23	98.90 ± 0.13	98.91 ± 0.14	97.88 ± 0.10	93.65 ± 0.39	93.61 ± 0.39	**99.59 ± 0.12**
	chess	1.04 ± 0.08	26.70 ± 2.03	78.81 ± 2.50	18.23 ± 0.41	1.12 ± 0.93	26.01 ± 1.39	**90.83 ± 0.02**
	isolet	80.51 ± 0.73	79.68 ± 1.10	80.39 ± 0.96	84.80 ± 1.03	86.00 ± 1.35	86.61 ± 1.26	**86.99 ± 0.54**
	Hill-with	89.09 ± 2.45	86.02 ± 2.22	88.85 ± 2.25	89.04 ± 2.95	89.83 ± 2.48	90.22 ± 2.76	**93.12 ± 4.20**
	Hill-without	99.96 ± 0.10	84.00 ± 1.51	81.78 ± 9.61	99.86 ± 0.10	99.68 ± 0.93	99.83 ± 0.22	**99.87 ± 0.19**
	musk	81.27 ± 1.30	85.48 ± 1.14	83.87 ± 1.43	82.90 ± 1.70	85.39 ± 1.43	84.35 ± 2.39	**88.63 ± 1.13**
	Sonar	83.46 ± 6.01	81.35 ± 6.57	82.50 ± 7.25	84.94 ± 5.33	86.06 ± 5.55	82.96 ± 5.84	**96.06 ± 1.93**
$cost_1 = 25$ $cost_2 = 10$	madelon	56.78 ± 0.48	56.59 ± 1.00	56.18 ± 1.18	56.46 ± 0.83	56.47 ± 1.12	55.35 ± 0.77	**61.35 ± 0.54**
	SECOM	98.86 ± 0.19	98.89 ± 0.22	98.93 ± 0.17	98.94 ± 0.22	99.86 ± 0.18	98.93 ± 0.17	**99.89 ± 0.06**
	chess	38.39 ± 0.30	**48.04 ± 0.43**	45.35 ± 0.46	44.26 ± 2.05	41.77 ± 1.22	46.97 ± 0.65	46.67 ± 1.12
	isolet	82.53 ± 0.44	78.44 ± 0.61	75.77 ± 6.46	80.76 ± 0.61	81.10 ± 0.64	82.04 ± 0.56	**83.17 ± 0.26**
	Hill-with	82.29 ± 3.21	87.89 ± 2.25	87.71 ± 3.06	87.65 ± 3.21	87.65 ± 3.21	91.62 ± 2.32	**92.89 ± 1.25**
	Hill-without	67.16 ± 1.32	36.61 ± 2.00	80.55 ± 2.20	77.77 ± 2.64	77.77 ± 2.64	79.31 ± 1.52	**82.46 ± 7.30**
	musk	81.02 ± 1.32	84.72 ± 1.63	83.57 ± 1.52	83.01 ± 1.22	84.35 ± 1.39	84.72 ± 1.86	**85.40 ± 1.88**
	Sonar	87.08 ± 3.28	87.10 ± 3.71	86.93 ± 3.70	87.06 ± 3.51	86.59 ± 3.66	85.67 ± 2.55	**87.23 ± 4.98**

The total cost, specificity and sensitivity are used as evaluation indicators to evaluate the performance of all methods on eight data sets.

The total cost is calculated as follows:

$$Total\ \ Cost = \mathrm{sum}\,(c_i)\,, \tag{20}$$

$$c_i = \begin{cases} cost_1 \quad \text{or} \quad cost_2 & ,\ \text{predicted\ \ label} \neq \text{true\ \ label} \\ 0 & ,\ \text{otherwise} \end{cases}, \tag{21}$$

where c_i represents the misclassification cost of a sample. If the predicted label is equal to the true label, the cost of c_i is 0, otherwise, the cost of c_i is equal

Table 4. The value of sensitivity on eight data sets. Bold numbers indicate the best results.

Cost	Data sets	CSLS	Semi-JMI	Semi-IMIM	CSEFS	CSFS	RLSR	Proposed
$cost_1 = 10$ $cost_2 = 25$	madelon	59.51 ± 0.68	59.93 ± 0.62	60.27 ± 0.54	55.42 ± 2.89	56.36 ± 1.22	55.32 ± 0.38	**61.56 ± 0.96**
	SECOM	80.46 ± 3.01	81.35 ± 3.57	82.50 ± 3.25	84.94 ± 4.33	84.06 ± 3.55	82.96 ± 4.84	**86.06 ± 0.30**
	chess	94.58 ± 0.13	**99.28 ± 0.10**	84.12 ± 1.65	94.52 ± 0.15	97.87 ± 0.21	97.44 ± 0.16	90.10 ± 0.02
	isolet	81.33 ± 0.76	78.89 ± 0.48	79.97 ± 0.81	84.95 ± 1.01	89.81 ± 0.62	90.03 ± 0.37	**92.11 ± 0.68**
	Hill-with	73.56 ± 1.68	77.81 ± 2.09	80.93 ± 1.06	77.01 ± 1.81	90.68 ± 2.82	90.92 ± 2.28	**91.84 ± 1.45**
	Hill-without	99.15 ± 0.18	92.31 ± 4.24	90.10 ± 4.75	99.46 ± 0.22	99.49 ± 0.42	99.10 ± 0.42	**99.58 ± 0.45**
	musk	81.19 ± 1.78	83.54 ± 1.58	81.54 ± 2.17	82.29 ± 1.33	83.22 ± 1.97	82.60 ± 2.69	**86.40 ± 1.88**
	Sonar	38.38 ± 12.61	38.61 ± 11.28	3.91 ± 12.62	38.07 ± 11.91	38.49 ± 12.74	36.63 ± 13.26	**58.10 ± 2.31**
$cost_1 = 25$ $cost_2 = 10$	madelon	56.36 ± 0.93	56.04 ± 0.89	56.11 ± 0.93	56.66 ± 1.45	57.30 ± 0.84	56.22 ± 1.08	**60.16 ± 0.54**
	SECOM	85.08 ± 3.02	86.10 ± 3.71	86.93 ± 2.70	28.63 ± 2.04	87.06 ± 3.51	83.59 ± 3.10	**87.22 ± 3.32**
	chess	98.25 ± 0.66	97.87 ± 0.25	98.35 ± 0.18	97.50 ± 0.88	98.78 ± 0.39	98.05 ± 0.22	**98.98 ± 0.02**
	isolet	89.02 ± 0.51	86.49 ± 0.62	87.37 ± 0.59	90.36 ± 0.45	90.61 ± 0.41	90.71 ± 0.30	**90.96 ± 0.18**
	Hill-with	67.16 ± 1.32	76.61 ± 2.00	80.55 ± 2.20	77.77 ± 2.64	74.77 ± 2.65	79.31 ± 1.52	**82.46 ± 7.30**
	Hill-without	90.91 ± 3.13	58.55 ± 9.90	56.17 ± 9.71	96.26 ± 3.41	99.71 ± 0.26	98.90 ± 0.25	**99.86 ± 0.12**
	musk	80.06 ± 1.50	83.00 ± 1.28	82.48 ± 1.28	82.29 ± 1.71	83.22 ± 1.97	80.87 ± 1.70	**84.23 ± 1.27**
	Sonar	39.46 ± 12.24	39.85 ± 12.75	39.21 ± 11.71	38.86 ± 11.06	38.37 ± 9.50	44.99 ± 8.24	**46.92 ± 11.92**

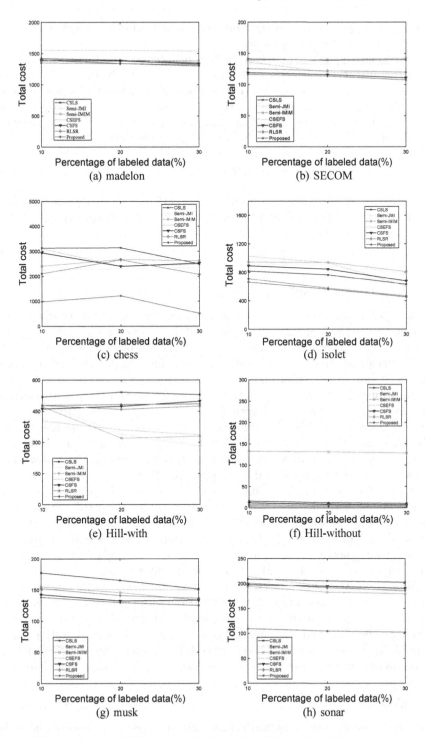

Fig. 1. The total cost of different methods under different labeled samples, at eight data sets while $cost_1 = 10$, $cost_2 = 25$.

to $cost_1$ or $cost_2$ ($cost_1$ and $cost_2$ represent the costs of being judged as positive and negative samples, respectively).

For a binary classification, there are four possible results: TP (True Positive) is positive instances correctly classified and TN (True Negative) is negative instances correctly classified. FP (False Positive) is negative instances incorrectly classified and FN (False Negative) is positive instances misclassified.

Specificity refers to the proportion of samples that are actually negative which are judged to be negative. It can be calculated by the following formula:

$$specificity = \frac{TN}{FP + TN} \tag{22}$$

Sensitivity refers to the proportion of samples that are actually positive which are judged to be positive. It can be calculated by the following formula:

$$sensitivity = \frac{TP}{TP + FN} \tag{23}$$

3.3   Experiment Results and Analysis

In this experiment, we reported the cost, specificity and sensitivity of all methods on eight UCI datasets in Table 2, Table 3 and Table 4 under different cost value settings and listed our observations as follows. In addition, we use a line chart to show the changing trend of the total cost under different proportions of labeled samples. It can be seen from Fig. 1.

From Table 2, we can know that the proposed SF_CSSI method outperformed other methods on most cases. Especially, on the chess data set, the total cost of SF_CSSI has reduced by 75% compared with the second best approach Semi-JMI, when $cost_1 = 10$ and $cost_2 = 25$. When $cost_1 = 25$ and $cost_2 = 10$, 31% reduction was achieved by the proposed method SF_CSSI on the Hill-without data set, compared to the second best approach CSFS.

From Table 3 and Table 4, the proposed model has high specificity and sensitivity. The highest specificity was obtained on SECOM and Hill-without data sets. The highest sensitivity was obtained on isolet and Hill-without data sets compared with other methods. In addition, specificity and sensitivity are commonly used diagnostic methods in clinical practice. The higher the value is, the more real, reliable and practical the diagnosis result will be.

From Fig. 1, the more labeled data we have, the lower cost we can achieve, in most cases. We also notice that SF_CSSI outperformed other CSLS, CSFS and CSEFS methods on almost all cases, which indicates that CSLS, CSFS and CSEFS can be improved with unlabeled data. This verifies the effectiveness of the semi-supervised feature selection method. In addition, the proposed method has the minimum total cost on most cases, especially in Hill-without data set.

3.4   Conclusion

This paper considers the misclassification and the structural information of the paired samples on each feature dimension. In addition, the information theory method is used to introduce a feature information matrix to simultaneously maximize joint relevancy of different pairwise feature combinations in relation to the target feature graphs and minimize redundancy among selected features. Compared with previous research on semi-supervised feature selection, this paper comprehensively considers the cost of misclassification, the structure information of paired samples on the feature dimension, and the information relationship of paired features. In general, it is more interpretable and generalizable for our method than others in this paper. Experiments on 8 real data sets show that the proposed method has better feature selection results.

In future work, we will try to extend our method to conduct a cost-sensitive multi-class classification.

Acknowledgment. This work was supported by the National Natural Science Foundation of China (Grant No: 81701780); the Research Fund of Guangxi Key Lab of Multi-source Information Mining and Security (No. 20-A-01-01); the Guangxi Natural Science Foundation (Grant No: 2017GXNSFBA198221); the Project of Guangxi Science and Technology (GuiKeAD20159041,GuiKeAD19110133); the Innovation Project of Guangxi Graduate Education (Grants No: JXXYYJSCXXM-011).

References

1. Zhang, S., Li, X., Zong, M., Zhu, X., Wang, R.: Efficient KNN classification with different numbers of nearest neighbors. IEEE Trans. Neural Netw. Learn. Syst. **29**(5), 1774–1785 (2017)
2. Gao, L., Guo, Z., Zhang, H., Xu, X., Shen, H.T.: Video captioning with attention-based LSTM and semantic consistency. IEEE Trans. Multimed. **19**(9), 2045–2055 (2017)
3. Shen, H.T., et al.: Heterogeneous data fusion for predicting mild cognitive impairment conversion. Inf. Fusion **66**, 54–63 (2021)
4. Zhu, X., Song, B., Shi, F., Chen, Y., Shen, D.: Joint prediction and time estimation of COVID-19 developing severe symptoms using chest CT scan. Med. Image Anal. **67**, 101824 (2021)
5. Lei, C., Zhu, X.: Unsupervised feature selection via local structure learning and sparse learning. Multimed. Tools Appl. **77**(22), 2960–2962 (2018)
6. Zhu, X., Zhang, S., Hu, R., Zhu, Y., Song, J.: Local and global structure preservation for robust unsupervised spectral feature selection. IEEE Trans. Knowl. Data Eng. **30**(99), 517–529 (2018)
7. Zhu, X., Li, X., Zhang, S.: Block-row sparse multiview multilabel learning for image classification. IEEE Trans. Cybern. **46**(46), 450 (2016)
8. Wu, X., Xu, X., Liu, J., Wang, H., Nie, F.: Supervised feature selection with orthogonal regression and feature weighting. IEEE Trans. Neural Netw. Learn. Syst. **99**, 1–8 (2020)
9. Zheng, W., Zhu, X., Wen, G., Zhu, Y., Yu, H., Gan, J.: Unsupervised feature selection by self-paced learning regularization. Pattern Recogn. Lett. **132**, 4–11 (2020)
10. Zhu, X., Zhang, S., Zhu, Y., Zhu, P., Gao, Y.: Unsupervised spectral feature selection with dynamic hyper-graph learning. IEEE Trans. Knowl. Data Eng. (2020). https://doi.org/10.1109/TKDE.2020.3017250

11. Shen, H.T., Zhu, Y., Zheng, W., Zhu, X.: Half-quadratic minimization for unsupervised feature selection on incomplete data. IEEE Trans. Neural Netw. Learn. Syst. (2020). https://doi.org/10.1109/TNNLS.2020.3009632

12. Cai, J., Luo, J., Wang, S., Yang, S.: Feature selection in machine learning: a new perspective. Neurocomputing **300**(jul.26), 70–79 (2018)

13. Shi, C., Duan, C., Gu, Z., Tian, Q., An, G., Zhao, R.: Semi-supervised feature selection analysis with structured multi-view sparse regularization. Neurocomputing **330**, 412–424 (2019)

14. Bennett, K.P., Demiriz, A.: Semi-supervised support vector machines. In: Advances in Neural Information Processing Systems, pp. 368–374 (1999)

15. Zhao, Z., Liu, H.: Semi-supervised feature selection via spectral analysis. In: Proceedings of the 2007 SIAM International Conference on Data Mining, pp. 641–646 (2007)

16. Ren, J. Qiu, Z., Fan, W., Cheng, H., Philip, S.Y.: Forward semi-supervised feature selection. In: Pacific-Asia Conference on Knowledge Discovery and Data Mining, pp. 970–976 (2008)

17. Chen, X., Yuan, G., Nie, F., Huang, J.Z.: Semi-supervised feature selection via rescaled linear regression. In: IJCAI, pp. 1525–1531 (2017)

18. Moosavi, M.R., Jahromi, M.Z., Ghodratnama, S., Taheri, M., Sadreddini, M.H.: A cost sensitive learning method to tune the nearest neighbour for intrusion detection. Iran. J. Sci. Technol. - Trans. Electr. Eng. **36**, 109–129 (2012)

19. Bai, L., Cui, L., Wang, Y., Yu, P.S., Hancock, E.R.: Fused lasso for feature selection using structural information. Trans. Knowl. Data Eng. 16–27 (2019)

20. Liu, M., Xu, C., Luo, Y., Xu, C., Wen, Y., Tao, D.: Cost-sensitive feature selection by optimizing F-measures. IEEE Trans. Image Process. **27**(3), 1323–1335 (2017)

21. Lin, J.: Divergence measures based on the shannon entropy. IEEE Trans. Inf. Theory **37**(1), 145–151 (1991)

22. Bai, L., Hancock, E.R.: Graph kernels from the Jensen-Shannon divergence. J. Math. Imaging Vis. **47**(1), 60–69 (2013)

23. Wang, H., et al.: Sparse multi-task regression and feature selection to identify brain imaging predictors for memory performance. In: 2011 International Conference on Computer Vision, pp. 557–562 (2011)

24. Miao, L., Liu, M., Zhang, D.: Cost-sensitive feature selection with application in software defect prediction. In: Proceedings of the 21st International Conference on Pattern Recognition (ICPR 2012), pp. 967–970 (2012)

25. Sechidis, K., Brown, G.: Simple strategies for semi-supervised feature selection. Mach. Learn. **107**(2), 357–395 (2018)

26. Zhao, H., Yu, S.: Cost-sensitive feature selection via the $l_{2,1}$-norm. Int. J. Approx. Reason. **104**(1), 25–37 (2019)

27. Melacci, S., Belkin, M.: Laplacian support vector machines trained in the primal. J. Mach. Learn. Res. **12**(3), 1149–1184 (2011)

Contextual Bandit Learning for Activity-Aware Things-of-Interest Recommendation in an Assisted Living Environment

May S. Altulayan[1], Chaoran Huang[1(✉)], Lina Yao[1(✉)], Xianzhi Wang[2(✉)], and Salil Kanhere[1(✉)]

[1] The University of New South Wales, Sydney, NSW 2052, Australia
{m.altulyan,chaoran.huang,lina.yao,salil.kanhere}@unsw.edu.au
[2] The University of Technology Sydney, Ultimo, NSW 2007, Australia
xianzhi.wang@uts.edu.au

Abstract. Recommendation systems are crucial for providing services to the elderly with Alzheimer's disease in IoT-based smart home environments. Therefore, we present a Reminder Care System to help Alzheimer patients live safely and independently in their homes. The proposed recommendation system is formulated based on a contextual bandit approach to tackle dynamicity in human activity patterns for accurate recommendations meeting user needs without their feedback. Our experiment results demonstrate the feasibility and effectiveness of the proposed Reminder Care System in real-world IoT-based smart home applications.

Keywords: Contextual bandit · IoT · Recommender system

1 Introduction

Alzheimer's disease (AD) is the most common type of dementia with severe implications on the day-to-day activities of numerous people world-wide [1]. In the US, 6.08 million elderly people reportedly suffer from clinical AD or mild cognitive impairment in 2017 with a potential escalation of this figure to 15.0 million by 2060 [2]. Meanwhile, the cost implication in the areas of providing care for Americans with AD and other dementia is equally as a major concern at an estimated \$290 billion in 2019 [3], opposed to the 2018 figure which stood at \$277 billion [4].

The various stages of AD are generally grouped into three, mild, moderate, and late or severe stages. Each of these stages presents different symptoms with the mild and moderate stages capable of lasting for about 3 years while the late or severe stage could last throughout the remainder of the patient's life. In the mild stage, patients begin to lose only short-term memory where they forget their ability to remember people's names or recent events. This stage is best manageable with technological aids. However, patients at moderate stage may

© Springer Nature Switzerland AG 2021
M. Qiao et al. (Eds.): ADC 2021, LNCS 12610, pp. 37–49, 2021.
https://doi.org/10.1007/978-3-030-69377-0_4

have acute memory loss which could affect the ability to handle some simple tasks, language problems, time consideration, and some changes in their personality which may become emotional. For the last stage, patients lose their ability to talk, understand, swallow, and walk. Consequently, intensive care from family members or professional caregivers is required [1].

Recommender systems could be adapted to help patients live safely and independently especially in IoT-based smart-home environments during the mild stage, and they are increasingly employed to provide critical services, e.g. reminder care services. A reminder care system is especially suitable for the mild stage, as at this stage, patients just begin to lose short-term memory (e.g. difficulty in remembering people's names and recent events) [1,5] without losing the ability to use such a system. A reminder care system in a smart-home environment would generally exploit sensory data from various sources e.g., environmental sensors, wearable sensors, and appliance sensors to deliver reminder recommendations to patients of items that they might need. This does not necessarily require feedback to improve the quality of recommendations. For instance, using the following scenario to demonstrate the importance of a reminder care system for Alzheimer patients: Aris is a 79-year-old woman living alone with mild AD. The reminder care system automatically monitors and recognizes Aris' activity patterns and recommends items that might be needed based on Aris' status and activity patterns. For example, if she completes cooking but forgets to turn off the stove, the system is set to remind her of this promptly. Afterwards, the system checks her acceptance for the recommendation automatically to improve the quality without needing Aris' explicit feedback.

Several studies have focused on reminder recommender systems aimed at providing assistance to elderly people diagnosed with AD. Oyeleke et al. [6] propose a recommendations system to monitor the daily indoor activities of seniors with mild cognitive impairment while Ahmed et al. [7] design a smart biomedical assisted system to assist Alzheimer patients. Several studies [8–11] use smartphone applications to provide care services to AD patients. The dynamicity in human activity patterns has not been given desired attention in most previous works, thus delivering low-quality recommendations. Another notable issue is the increased focus on monitoring which gives a reminder to patients while the system has to wait for patients' feedback to update itself. From the illustration of Aris scenario above, suppose Aris has the following pattern when preparing a cup of coffee in the morning: (1) turning on the coffee machine, (2) bringing a cup, (3) putting some milk, and (4) then adding sugar. Then, if she brings a cup and forgets it, what is next? the system should remind her of grabbing the cup. However, if one day, she changes this pattern and decides not to add milk to her coffee in the future, the system should also cope with that. From Aris' standpoint, she expects the system to be a caregiver, which helps only when needed without actively requesting feedback. Therefore, the system should be capable of assessing the quality of recommendations without requiring user feedback.

In our previous work [12,13], we developed a prototype system capable of detecting complex activities to enable the system to cope with the dynamicity in human activity patterns. Here, we extend the previous work by developing a recommender system that can not only learn the dynamicity of human activity pattern, but also remind patients about the correct item when the patients need it without requiring their feedback. To this end, we formulate our problem using a contextual bandit approach, which focuses on context as input to produce the next action. The main contributions of this paper are as follows:

- We propose a recommender system based on contextual bandit by fusing context information from past activities, current activity, and items to recommend the correct item.
- We update our system automatically without needing feedback from users to improve the recommendations.
- We evaluate the model using a public dataset and our experimental results demonstrate the feasibility and effectiveness of our approach.

2 Related Work

2.1 Recommender System for the IoT

Generally, recommendation systems can be used to assist users in selecting their preferences of items in IoT enriched environments. Some of these works exploit the traditional recommender system approaches: collaborative filtering [14], content-based [15] and hybrid-based approach [16] to build their systems. Authors in [17] propose a unified collaborative filtering model based on probabilistic matrix factorization recommender system that exploits three kinds of relations to extract the latent factors among these relations. In [15], the authors adopt a content-based solution for the recommender engine in their AGILE project which aims to improve the health conditions of users. Authors in [16,18,19] built their recommender system engine using a hybrid recommendation algorithm.

Reinforcement learning approach (RL) has also been adapted in building recommender system for the IoT environments. RL deals with dynamic environments and learns a policy that maximizes the long-term reward particularly for continuous record update. Massimo et al. [20,21] adopt inverse reinforcement learning to model user behaviours. Oyeleke et al. [6] design a system that monitors daily indoor activities for people with mild cognitive impairment. Most RL algorithms particularly dealing with dynamic environment focus on matching each state for an action using different policies sequentially. This is achieved by observing how the taken action could affect the next state by considering future rewards. However, RL cannot handle a system that needs to learn the best action in different scenarios and treat each state independently while not allowing one action to affect the next stat. Consequently, we formulate our recommender system with a contextual bandit approach.

2.2   Contextual Bandit Approach for Recommendation

Unlike traditional online machine learning algorithms that learn from offline data, contextual bandit (CB) approaches exploit both offline data and environment interactions. Contextual bandit takes common features from RL of using policy to take an action based on the context of each state, and similar to multi-armed bandit (MAB), it focuses on the immediate reward. Some studies have adapted CB for their recommender systems.

Li et al. [22] adopt contextual bandit for news article recommendation. The proposed algorithm, LinUCB, shows the ability to deal with sparse and large data combined with other algorithms such as ϵ-greedy. In [23], CB is adapted to build an online learning recommender system where information for history students learning and current student are used as context to conduct the learning recommendations to the student. Zhang et al. [24] propose a novel contextual bandit method named SAOR for online recommendations. It deals with sparse interactions by distinguishing between negative response and non-response to improve recommendation quality. We adopt CB approach to formulate the problem tackling two main challenges in our system, the dynamicity of human activity pattern to recommend the correct item and then knowing the feedback automatically without waiting for feedback from the user.

3   Contextual-Bandit-Based Reminder Care System

Our proposed system for reminder recommendation has three major stages (see Fig. 1): (1) Complex activity recognition stage, where we exploited three data sources, wearable sensors, environmental sensory data, and the usage of home appliances; (2) Prompt detection stage, which determines if an ongoing activity requires an item recommendation using data mining approaches; (3) reminder

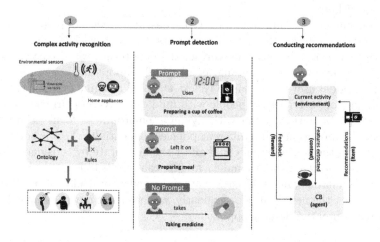

Fig. 1. Overview of the proposed methodology.

recommendation, which uses contextual bandit approach to extract context from the previous two stages and recommend items to the user during an activity. We further discuss the stages in the subsections below.

3.1 Complex Activity Detection

A system needs to detect what activity a user is performing before it can recommend an appropriate item to the user. Although Human Activity Recognition (HAR) has been studied extensively, most existing studies focus on detecting simple activities using wearable sensors, which are inadequate to support the detection of complex activities. Exploiting further sources, such as environmental sensors and home appliance sensors, helps the system to detect the complex activity accurately.

In our previous work [12], we designed a preliminary reminder care system that provides reminder recommendations based on complex activities detection. We conduct recommendations via three main steps:

– Elementary activity recognition. We use the common configuration of Deep-ConvLSTM as the classifier to detect elementary activities. DeepConvLSTM has 4 convolutional layers with feature maps and 2 LSTM layers with 128 cells. The result shows that DeepConvLSTM archives a promising accuracy of 77.2%.
– Ontology for complex activity recognition. After detecting elementary activities, we build an OWL (Ontology Web Language3) ontological models, which include the artefacts, environment, locations, and activities required to define things involved in the interaction.
– Rule-based orchestration. This step uses the output from the two previous steps to detect complex activities. It consists of a set of rules that are produced based on the previous ontological models.

3.2 Prompt Detection

This part uses the collected data from the previous stage to identify if an activity needs a prompt or not. A prompt is defined in two main situations: (1) when the user has been stuck within an activity for some time without taking an action; (2) when the user uses a wrong item that does not belong to this activity. Various learning models can be adopted at this stage to determine when the user needs a prompt during her activities. For example, Das et al. [25] test several classification methodologies on the PUCK dataset, including Support Vector Machines (SVM) [26], Decision Tree [27] and Boosting [28]. In particular, Boosting applies a classification algorithm to re-weight the training data versions sequentially and then extracted a weighted majority vote of the previous sequentially classifiers. And it generally outperforms the other two methods.

3.3   Conducting Recommendations

When the system is defined that the user's activity needs a prompt, the system at this stage should decide which item is suitable to be recommended at this moment based on the user situation. One of the main challenges, as we mentioned above, is that each activity could be done differently. The system has to consider which is a correct item to be recommended even it is the same activity by considering the user situation. This stage represents our main contributions in this paper.

Problem Definition. When a complex activity that needs a prompt is received by the agent G at time t, our algorithm extracts the context x and nominates an appropriate item a for the current activity. Then, the agent receives feedback as reward r for the recommended item. Finally, the system is being updated based on the received reward.

Algorithm 1: Our procedure to recommended a correct item for user's activity. It takes context x as input, and returns a recommended item as output a

---

Input: x
Output: a
   Procedure $agentrecommend(a)$
1: **for** $x_t \in X$ **do**
2:    $x_t \leftarrow PAC, CAC, IC$
3:    $a \leftarrow x_t$ agent G uses a policy to match the context for a correct item
4:    waiting for T_r
5:    compute $V(x)$
6:    put x_t, r_t, a into experience pool
7:    update the system
8: **end for**
9: **return**

Method. We formulate our problem as a contextual bandit approach to tackle the dynamicity of human activity patterns and to recommend the correct item without having to wait for the user's feedback. Contextual bandit provides a learning model based on context. Three kinds of context are extracted at this stage:

– **Past activities context (PAC).** Since each activity can have a different pattern, for each activity, the system extracts the path/sequences of items used in the past (recorded in the log file) as a type of context. We use the recorded paths of each activity as an experience pool based on which the agent can decide which item to recommend at a specific state.

- **Current activity Context (CAC).** When the system receives data from the previous two stages, it extracts the context about the current state. For example, when the system receives that the user needs a prompt for preparing coffee, the context of the current activity (locations, previous items, user position, and time.) will be extracted.
- **Item context (IC).** Item context includes information about items, such as to which activity this item belongs, how long could it be in use, and how many times the user needs it for the current activity. For example, a coffee machine as an item can be used for the activity of 'preparing coffee', where it can be being used for around 2 min each time.

The agent receives the above contextual information as input (see Algorithm 1). The contextual bandit combined three main components: an environment, which represents the context of the user's activity $x \in X$, an agent G, which chooses an action $a \in A$ (notice that the common name in CB is Learner but we call it an agent in our case) based on the received context, and a reward $r \in \{0,1\}$, which the agent aims to maximize by recommending the correct action at each round $t = 1, 2, ..., T$. We calculate the expected reward of each policy using the following equation:

$$V(x) = \frac{1}{T} \sum_{t=1}^{T} E[r_t | x, \pi(x)] \tag{1}$$

where the agent G can choose from a set of policies $\prod \subseteq \{x \to A\}$ by employing two streaming models: Linear regression and stochastic gradient distance (to be detailed in Sect. 4.3).

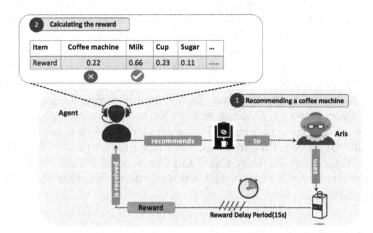

Fig. 2. (1) The agent recommends a coffee machine to Aris whereas she uses milk instead; then the system waits for 15 s; (2) The feedback is received by the system as a reward and it is calculated accordingly as the coffee machine is the wrong item.

Most traditional recommender systems focus on 'click' or 'not click' as feedback to calculate the reward function immediately and to update the system. In contrast, our system recommends an item to the user and then waits for sufficient time to decide if the recommended item is used or not—by checking its status (on/off or moved/not moved). For example, if the system recommends a coffee machine to Aris (see Fig. 2) when she is preparing a cup of coffee, whereas she wants to use it later yet not immediately. This does not mean that the recommended item is incorrect, and it is better for the system to ignore this false negative feedback this time. To facilitate the above, we introduce a Reward Delay Period T_r, which accounts for the different paces of users in carrying out activities and is calculated as:

$$T_r = x_t + W_t \tag{2}$$

where x_t represents the user's activity and W_t represents the waiting time period. We consider W_t a hyperparameter (to be detailed Sect. 4.3).

4   Evaluation

We report our evaluation of the proposed system. by introducing the dataset that we fit into our system, the feature engineering process, and finally, our experimental results.

4.1   Dataset

We evaluate the proposed system on, PUCK [25], a public dataset published in 2011. The PUCK dataset collected from a Kyoto smart home testbed located in Washington State University in two-story apartments with one living room, one dining area, and one kitchen on the first floor and, one bathroom and three bedrooms on the second floor. It combines three types of sensory data: (1) environmental sensors, including motion sensors on ceilings, door sensors on room entrances, kitchen cabinet doors, microwave, and refrigerator doors, temperature sensors in rooms, power meter, burner sensor, water usage sensors, and telephone usage sensors, (2) items sensors for usage monitoring, and (3) two wearable sensors. Eight complex activities are defined: Sweep and Dust, DVD Selection and Operation, Prepare Meal, Fill Medication Dispenser, Water Plants, Outfit Selection, Write Birthday Card, and Converse on Phone. Also, activities are divided into ordered steps, which can help detect whether the activity is completed correctly.

4.2   Features Engineering

The PUCK dataset has four fields (date, time, sensor ID, and sensor value). To adapt the PUCK dataset for our system, we process it to extract the required features via the following step:

Table 1. Tuning hyperparameters for the OSL model policies

Policy	Note	Hyberparameters						
		beta_prior	alpha	Smoothing	Decay	refit_buffer	active_choice	decay_type
LinUCB [22]	LinUCB policy stores a square matrix which has dimension equal to total numbers of features for the fitted model	None	0.1					
AdaptiveGreedy [29]	It focuses on taking the action that has the highest reward	None		(1,2)	0.9997			Percentile
AdaptiveGreedy (Active)	It is the same for AdaptiveGreedy but with different hyberparameters	((3./nchoices, 4), 2)		None	0.9997		Weighted	Percentile
SoftmaxExplorer [29]	It depends on softmax function to select the action	None		(1,2)		50		
EpsilonGreedy [29]	It focuses on two way for taking the action: random action or the action with highest reward	None		(1,2)	None			
ActiveExplorer [29]	It depends on an active learning heuristic for taking the action	((3./nchoices, 4), 2)		None		50		

- Combining the environmental data sensors(motion, items, power/ burner/ water usage, door...etc.) with the wearable sensors for each participant.
- Labeling the complex activities for the whole dataset.
- Extracting the start and the end of each activity as a session to define when the user needs a prompt.
- Selecting only the common sensors among all participants, where the total measurement counts of each sensor greater than 25% for all the participants.
- Dividing the sensors into four groups (movement sensors, motion sensors, count sensors and, continuous values sensors) based on kind of measurements (e.g. binary value, continues value and multiple values)
  and then processing each group as follows:
  1. For the movement sensors group, extracting the following features: Mean, STD, Correlations.
  2. For motions sensors group, computing the fraction counts across the groups.
  3. Counting sensors that have on/off measurements.
  4. For the last group, calculating the average for continuous value sensors.
- After extracting all features, applying the previous group process for all the participant sessions.

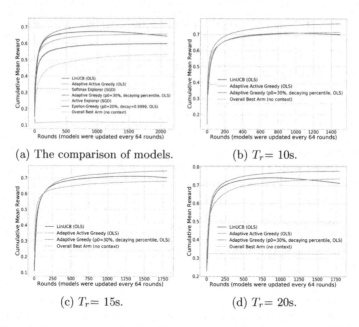

(a) The comparison of models.   (b) $T_r = 10s$.

(c) $T_r = 15s$.   (d) $T_r = 20s$.

Fig. 3. Cumulative mean reward with the comparison of online contextual bandit models (Streaming data mode, see Fig. a) and selected models with different Reward Delay Periods T_r (see Fig. b, c, d).

We take two methods to overcome the item usage imbalance problem (i.e, only a small number of items are frequently used): 1) Dropping outliers in items. This method is simple yet effective in improving the performance; 2) Sampling other random points in activity sessions to increase the prompt points, although this does not help balance the item usages as (1).

4.3 Experiment Results

We first evaluate the effectiveness of the contextual bandit approach in recommending the correct item to a user in case the user's current activity needs a prompt. The system uses all the extracted features as context to make a recommendation of the correct item. We use one public available contextual bandit package of python for our experiments. The package provides two types of models: full batch models and streaming models. Because of the sample limitation of the PUCK dataset, we focus on the streaming models, namely SGDClassifier (SGD) and LinearRegression.

Both models are sensitive to hyperparameters such as beta_prior or smoothing. However, SGDClassifier has stochastic matrices while LinearRegression (OLS) has matrices which are closed to the solution, and it updates them incrementally. Consequently, Linear regression performed better across different policies than SGD in our system (see Fig. 3a).

As shown in Fig. 3a, a set of policies are used for each model. Based on the results, we exclude SGD for the rest of our experiments due to its unpromising result. Details about parameters can be found in Table 1.

The Reward Delay Period T_r, as we mentioned above, plays the main role in defining when the agent receives the reward as a feedback of the recommended item. Tuning this parameter is important, as decreasing T_r could consider that the recommended item is not used while increasing T_r could confuse the agent specifically when the user starts to use other items before receives the feedback about the recommended one. The results in Fig. 3 show that when the T_r has a large value, it improves the accuracy to become around 0.78. Here, we treat T_r as a hyperparameter that can be adjusted based on each item; we will leave it to our future work. In addition, we can see here the system does not need to any feedback from the user to receives the reward. Consequently, it is calculated automatically after the Reward Delay Period. We focus on this feature because our system deals Alzheimer's patients which is difficult for them holding a smartphone and confirm their response for recommendations.

5 Conclusion

In this work, we explore the feasibility of building a system that makes reminder recommendations to Alzheimer's patients only when they need a reminder. We take advantage of the contextual bandit approach to formulate our problem and tackle two main issues: dynamicity of human activity pattern and recommending the correct item without needing explicit user feedback. Experiments demonstrate the effectiveness of our recommender system. One limitation lies in our evaluation of the system is that our experiments are still not comprehensive enough because the only suitable dataset that we use does not include time labels, which are, however, one important and critical type of context. In the future, we will create our own test-bed to collect inclusive and adequate data for complex experiments, and testing our framework in real-life scenarios.

References

1. Alzheimer's society. Accessed 07 Jan 2019
2. Brookmeyer, R., Abdalla, N., Kawas, C.H., Corrada, M.M.: Forecasting the prevalence of preclinical and clinical alzheimer's disease in the united states. Alzheimer's Dementia **14**(2), 121–129 (2018)
3. Association, A.: 2019 alzheimer's disease facts and figures. Alzheimer's Dementia **15**(3), 321–387 (2019)
4. Association, A.: 2018 alzheimer's disease facts and figures. Alzheimer's Dementia **14**(3), 367–429 (2018)
5. Yao, L., Wang, X., Sheng, Q.Z., Dustdar, S., Zhang, S.: Recommendations on the internet of things: requirements, challenges, and directions. IEEE Internet Comput. **23**(3), 46–54 (2019)
6. Oyeleke, R., Yu, C., Chang, C.: Situ-centric reinforcement learning for recommendation of tasks in activities of daily living in smart homes. In: 42nd Annual Computer Software and Applications Conference, vol. 2, pp. 317–322. IEEE (2018)

7. Ahmed, Q.A., Al-Neami, A.Q.: A smart biomedical assisted system for alzheimer patients. In: IOP Conference Series: Materials Science and Engineering, vol. 881, p. 012110. IOP Publishing (2020)

8. Armstrong, N., Nugent, C., Moore, G., Finlay, D.: Developing smartphone applications for people with alzheimer's disease. In: The 10th International Conference on Information Technology and Applications in Biomedicine, pp. 1–5. IEEE (2010)

9. Choon, L.: Helper system for managing alzheimer's people using mobile application. Universiti Malaysia Pahang (2015)

10. Alharbi, S., Altamimi, A., Al-Qahtani, F., et al.: Analyzing and implementing a mobile reminder system for alzheimer's patients, pp. 444–454 (2019)

11. Aljehani, S., Alhazmi, R., Aloufi, S., Aljehani, B., Abdulrahman, R.: iCare: applying IoT technology for monitoring alzheimer's patients. In: 1st International Conference on Computer Applications & Information Security. IEEE (2018)

12. Altulyan, M.S., Huang, C., Yao, L., Wang, X., Kanhere, S., Cao, Y.: Reminder care system: an activity-aware cross-device recommendation system. In: Li, J., Qin, S., Li, X., Wang, S., Wang, S. (eds.) ADMA 2019. LNCS (LNAI), vol. 11888, pp. 207–220. Springer, Cham (2019). https://doi.org/10.1007/978-3-030-35231-8_15

13. Yao, L., et al.: WITS: an IoT-endowed computational framework for activity recognition in personalized smart homes. Computing **100**(4), 369–385 (2018)

14. Sarwar, B., Karypis, G., Konstan, J., Riedl, J.: Item-based collaborative filtering recommendation algorithms. In: Proceedings of the 10th International Conference on World Wide Web, pp. 285–295 (2001)

15. Pazzani, M.J., Billsus, D.: Content-based recommendation systems. In: Brusilovsky, P., Kobsa, A., Nejdl, W. (eds.) The Adaptive Web. LNCS, vol. 4321, pp. 325–341. Springer, Heidelberg (2007). https://doi.org/10.1007/978-3-540-72079-9_10

16. De Campos, L., Fernández-Luna, J., Huete, J., et al.: Combining content-based and collaborative recommendations: a hybrid approach based on bayesian networks. Int. J. Approx. Reason. **51**(7), 785–799 (2010)

17. Yao, L., Sheng, Q.Z., Ngu, A.H., Ashman, H., Li, X.: Exploring recommendations in internet of things. In: the 37th international ACM SIGIR Conference on Research & Development in Information Retrieval, pp. 855–858. ACM (2014)

18. HamlAbadi, K., Saghiri, A., Vahdati, M., et al.: A framework for cognitive recommender systems in the internet of things. In: The 4th International Conference on Knowledge-Based Engineering and Innovation, pp. 0971–0976. IEEE (2017)

19. Saghiri, A., Vahdati, M., Gholizadeh, K., et al.: A framework for cognitive internet of things based on blockchain. In: The 4th International Conference on Web Research, pp. 138–143. IEEE (2018)

20. Massimo, D.: User preference modeling and exploitation in IoT scenarios. In: 23rd International Conference on Intelligent User Interfaces, pp. 675–676 (2018)

21. Massimo, D., Elahi, M., Ricci, F.: Learning user preferences by observing user-items interactions in an IoT augmented space. In: Adjunct Publication of the 25th Conference on User Modeling, Adaptation and Personalization. ACM (2017)

22. Li, L., Chu, W., Langford, J., Schapire, R.: A contextual-bandit approach to personalized news article recommendation. In: Proceedings of the 19th International Conference on World Wide Web, pp. 661–670 (2010)

23. Intayoad, W., Kamyod, C., Temdee, P.: Reinforcement learning based on contextual bandits for personalized online learning recommendation systems. Wirel. Pers. Commun. 1–16 (2020)

24. Zhang, C., Wang, H., Yang, S., Gao, Y.: A contextual bandit approach to personalized online recommendation via sparse interactions. In: Yang, Q., Zhou, Z.-H., Gong, Z., Zhang, M.-L., Huang, S.-J. (eds.) PAKDD 2019. LNCS (LNAI), vol. 11440, pp. 394–406. Springer, Cham (2019). https://doi.org/10.1007/978-3-030-16145-3_31
25. Das, B., Cook, D., Schmitter-Edgecombe, M., Seelye, A.: Puck: an automated prompting system for smart environments: toward achieving automated prompting–challenges involved. Pers. Ubiquit. Comput. **16**(7), 859–873 (2012)
26. Boser, B., Guyon, I., Vapnik, V.: A training algorithm for optimal margin classifiers. In: The 5th Annual Workshop on Computational Learning Theory, pp. 144–152 (1992)
27. Quinlan, J.: Induction of decision trees. Mach. Learn. **1**(1), 81–106 (1986)
28. Friedman, J., Hastie, T., Tibshirani, R., et al.: Additive logistic regression: a statistical view of boosting (with discussion and a rejoinder by the authors). Ann. Stat. **28**(2), 337–407 (2000)
29. Cortes, D.: Adapting multi-armed bandits policies to contextual bandits scenarios. arXiv preprint arXiv:1811.04383 (2018)

Deep Multi-view Spatio-Temporal Network for Urban Crime Prediction

Usama Salama[1], Xiaocong Chen[1(✉)], Lina Yao[1], Hye-Young Paik[1], and Xianzhi Wang[2]

[1] University of New South Wales, Sydney, NSW 2052, Australia
{u.salama,xiaocong.chen,lina.yao,h.paik}@unsw.edu.au
[2] University of Technology Sydney, Ultimo, NSW 2007, Australia
xianzhi.wang@uts.edu.au

Abstract. Crimes sabotage various societal aspects, such as social stability, public safety, economic development, and individuals' quality of life. To accurately predict crime occurrences can not only bring the peace of mind to individuals but also help distribute and manage police resources effectively by authorities. We aim to take into account plenty of environmental factors, such as data collected from Internet of Things (IoT) devices and social networks to predict crimes at city or a finer level. To this end, we propose a deep-learning-based spatio-temporal multi-view model, which explores the relationship between tweets, weather (a type of sensory data) and crime rate, for effective crime prediction. Our extensive experiments on a four-month crime dataset (covering 77 communities, 22 crime types, and 120 days) of Chicago city show that our model can achieve improvement over 19 out of 22 crime types (up to 6.7% in homicide). We also collect the corresponding weather information for different regions of Chicago city to support the crime prediction. Our experiments demonstrate that weather information can improve the performance of the proposed method.

Keywords: Spatial-temporal learning · Deep neural networks · Crime prediction

1 Introduction

Crime rates are reportedly rising continuously in several countries and regions. High crime rates undermine a society in multiple aspects, such as public safety, urban stability, medical expenditure, and economic development. It has, therefore, become a key issue to predict crime hotspots (e.g., the places and days when the crime rate would likely to go up) as a key effort on preventing crime effectively. To predict crime occurrences accurately will not only provide timely information for the general public but also assist in allocating police resources more efficiently.

U. Salama and X. Chen—Equal contribution.

© Springer Nature Switzerland AG 2021
M. Qiao et al. (Eds.): ADC 2021, LNCS 12610, pp. 50–61, 2021.
https://doi.org/10.1007/978-3-030-69377-0_5

Several research efforts have contributed to crime prediction. Wang et al. [19] propose a negative binomial regression model to infer a certain community's crime rate based on the characteristics of this community as well as the crime rates in other communities. They introduce two new types of data (Points of Interest and Taxi flows) and explore the correlation the relationship between these two features and crime rate. However, this work estimates crime rate using statistical crime records from other communities in the same year and thus cannot forecast crimes. Gerber et al. [9] combine twitter features and traditional kernel density estimate (KDE) for crime prediction. They employ logistic regression to utilize these features for crime probability estimate and apply distance-weighted spatial interpolation to smooth boundary. Their experiments confirm twitter information can improve the prediction performance in most crime types (19 out of 22).

In light that various features aside from topics (e.g., sentiment index) can be extracted from twitter and can play an important role in crime prediction, Al et al. [1] improve KDE by utilizing of spatio-temporally tagged Twitter posts for inferring micro-level movement patterns. They add temporal information and routine activity patterns inferred from social media as an supplementary feature for crime prediction. Spatio-temporal information is used on other prediction tasks as well [2,17]. Zhao et al. [22] propose a framework TCP to capture temporal-spatial correlation in urban data for crime prediction. They collect a series of statistical information (e.g., historical crime records, check-ins, point-of-interests) as features and further model intra-region temporal correlation and inter-region spatial correlation as regularization for the loss function. Meanwhile, previous research has shown weather influences violent behavior, e.g., a higher temperature generally leads to a higher crime rate [13]. Although previous studies have tried a series of methods to improve crime prediction performance, there is still a large space for improvement:

- Twitter has shown a great potential for event prediction [15]. However, instead of simply using topical and temporal features, a set of context-specific high quality features should be exploited from tweets.
- Deep learning models have demonstrated capability of learning the complex relationships from multi-source data in a variety of applications. It can be adapted for exploiting latent relationships and patterns from different sources of data.
- Crime occurrences might be sparse for certain crime types, resulting in an imbalanced dataset. Hence, a methodology is required to alleviate this problem to sustain an accurate prediction.
- Taxi-flow and geographical distance between different regions can provide a useful spatial correlation to help predict crime, which, however, it yet to be utilized in an efficient manner.
- How the magnitude and spatial distribution of criminal activity is affected by climatic conditions remains largely unexplored in the literature.

Targeting at the above points, we propose a spatial-temporal multi-view model for crime prediction, which utilizes deep learning models to exploit latent correlations and patterns from social and statistical information while simultaneously taking both temporal and spatial relationship into consideration. In summuary, we make the following main contributions in this paper:

- We present a unified model that simultaneously exploit latent relationships from the social information (tweets), statistical information (crime records), and weather information (rain or not) for crime prediction. In particular, we extract a set of high-quality tweet-specific features (emotional, criminal-related, and topical features) for the prediction.
- We efficiently utilize spatial relationships (taxi-flow and geographical distance) between different regions in our model by transforming them into regularization in the loss function. Besides, we utilizes GANs to generate real-time weather information to help predict crimes.
- We propose an iterative Smote-Tomeklinks method, which effectively mitigate the class imbalance problem caused from sparse occurrences in several crime types. Our extensive experiments in real dataset shows that our model outperforms a series of baseline and state-of-the-art methods.

2   Problem Formulation

We aim to predict crime occurrences in Chicago. To this end, we first partition the entire city into different regions. There are two common region-partition methods in previous studies: some [1,9,22] manually divide the entire city into disjoint regions with various grid sizes (e.g., 2 km × 2 km); others [19] use well-defined, historically recognized community areas. In our work, we divided the city of Chicago to eight regions based on eight main climate stations for Chicago city which are located in Aurora Municipal airport, Botancical garden, Midway airport, Northerly island, Ohare international airport, Palwaukee airport, Waukegan regional airport, and West Chicago Dupage airport. We used coordinates to scope region partition to guarantee weather data for the region is the closest to the data collected from each climate station. After region partition, we used coordinates to divide crime data to similar regions. We then matched the crime location using coordinates of each weather station region.

We analyze the statistical crime data of Chicago and find that a certain crime type generally occurs 1 or 0 times in a certain community on a specific day, with a very low possibility of appearing multiple times. In this context, we transform crime prediction problem into a binary-classification task (happen or not happen). Specifically, we denote by $X_{i_k} \in R^{T_1}$ the historical observed records of a crime type k with time window T_1, $T_i \in R^{T_2}$ the features extracted from tweets posted in community i during time window T_2, W_k^G the geographic distances between community i and the other 76 communities, and finally, W_k^T the taxi flow from the other 76 communities.

3   Data Collection and Feature Extraction

In the past few years, the American government has tried to record crimes precisely to help law enforcement analyze crime patterns and forecast crime hotspots in various cities. Among all the cities in the US, Chicago is ranked third in population (2.7 million) with a high crime rate in serious offenses, such as murders, robberies, aggravated assaults, and property crimes. The 'Chicago Data Portal'[1] provides a complete listing of data categories about this city including taxi trips, parks and recreations, and education, including detailed and immediate-update criminal records. Similar to [9], we collected information on all crimes ($n = 374,043$) documented for the whole year of 2011. Each crime record contains important information, such as criminal type (1 out of 30 types, eg. theft, battery and assult), date, description, location (latitude, logitude), and location description. After data cleaning and preparation, we calculate crimes corresponding occurrence frequencies, as shown in Table 1. According to our statistical results, the occurrence frequencies of different crime types fluctuate within an wide range (from 4 to 76145).

Table 1. Frequency of crime types in Chicago documented for the year 2011. We exclude the crime types with very low frequencies.

Crime types	No.	Crime types	No.
Theft	76145	Battery	72841
Criminal damage	43341	Deceptive practice	11331
Assault	24326	Other offense	21844
Narcotics	37264	Burglary	20540
Vehicle theft	21278	Robbery	14339
Criminal trespass	9722	Weapons violation	3248
Children offense	1671	Sexual assault	1319
Public peace violation	2127	Public officer interference	320
Sex offense	1667	Prostitution	4595
Homicide	3751	Arson	786
Liquor law violation	1284	Kidnapping	727
Gambling	810	Intimidation	221

Features From Social Media. We extract three types of features from tweets posted for each training window: emotional features, crime-related features, and topic features. Previous study [9] has shown the importance of topic features from tweets for crime prediction. Besides, we believe other features mined from the tweet content may also play an important role, e.g., the emotional and sentiment index from tweets in a certain community can provide strong supplementary information for crime prediction. We describe each feature type, respectively, as follows: 1) We analyze Emotional Features (F_{E_M}) in terms of emotional states, emotional intensity, sentiment score and content polarity. We measure emotional

---

[1] https://data.cityofchicago.org.

states using 10 emotional terms in tweets, such as 'hate', 'joy', 'sadness', 'anger', and 'nervousness'. Each term is estimated a score using the Empath API [8]. We measure Polarity as a score within the range of (0, 1) by analyzing the whole tweet using the Textblob API. Additionally, we also measure the positive and negative index (ranging from 0 to 1). 2) Regarding crime-related features, we take crime-related indexes from tweets into consideration. We choose up to 12 indexes, such as 'violence', 'aggression', 'dispute', 'Swearing Terms', 'fight', and 'weapon'. Similar to emotional features, these crime indexes also range from 0 to 1 with higher values indicating stronger relationships. 3) Regarding Topical Features F_T, we assume the topic distribution is consistent within each community and extract the topic distribution from a pseudo-document formed by all the filtered tweets from each community during a time window. We firstly exclude non-English tweets and tokenize each tweet using TweetTokenizer[2]. Then, we remove stop words and low frequency words. Differing from previous work [9], we use the original latent Dirichlet Allocation (LDA) to extract topics over unigrams and biterm topic model (BTM) [20] to extract topics over biterms, respectively. We set the number of topics for both topic model to 15. After combining the unigram and biterm topic features together, we get a 30 dimensinal topical feature vector. By assuming that the unigram topical feature and bigram topical feature can be generated from the same hidden representation, we apply a robust auto-encoder [22] to compress the joint topical feature into 15 dimensions. Auto-encoders are widely used to learn latent representation in several research works [5,14]. Finally, we get a 40-dimensional feature vector after feature extraction and partial feature transformation.

Features from Previous Criminal Record Spatial Correlations between Communities. Previous studies have shown that the crime rate at one location is highly correlated with nearby locations [11,19]. Therefore, we take geographical relationships into consideration in crime prediction. Different from [19], we form a geo-matrix $W^G(77 \times 77)$ by setting $W_{ii}^G = 0$ and W_{ij}^G as the euclidean distance between the centroids of communities i and j (W_{ij}^G is symmetric)[3]. Then, all the elements W_{ij}^G of the matrix are normalized over the row W_i^G. Besides geographical influence, taxi flows can also build a bridge between different communities and influence the crime rate [12,19]. Taking this into account, we build a taxi-flow matrix W^T in a similar manner as W^G, where each W_{ii}^T is set as zero and W_{ij}^T (W_{ij}^T is asymmetric) is calculated by summing up all the taxi flows from community j to i.

Over-Sampling and Data Cleaning. Before feeding the extracted features into our model, we conduct data interpolation in the minority class of data to solve the imbalance problem. Previous study [3] has conducted rich analyses of existing over-sampling methods and propose a new variant Smote+Tomek links, which shows excellent performance in their experiments. The original Smote-Tomeklinks successfully combine the advantages of Smote (over-sampling) [4]

---

[2] https://www.nltk.org/api/nltk.tokenize.html.
[3] https://github.com/brandonxiang/geojson-python-utils.

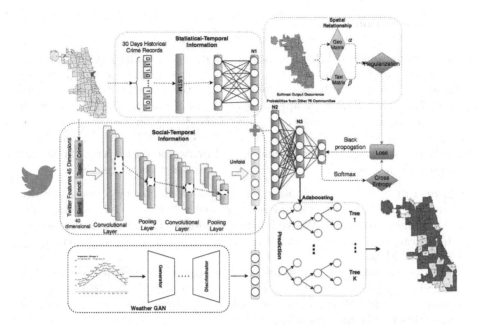

Fig. 1. The structure of the proposed model which utilize three different types of data: crime record, twitter and weather data.

and Tomek-links [18] (data-cleaning)—the method first oversamples minority data samples with Smote and then identifies and removes Tomek links. Details of Smote and Tomeklinks can be found in [4,18]. However, this method still has an disadvantage if a serous imbalance exists in a dataset because the noises in the dataset can actually influence the final calssification performance. More specifically, Smote might create a large amount of artificially invented samples, which contain considerable noises that cannot be removed by Tomeklinks. In this regard, we proposed an iterative Smote-Tomeklinks model, which employ Smote and Tomeklinks iteratively to clean out the noises in a timely manner. The overall process is shown in Fig. 2. The key difference of our method from the original Smote+Tomeklinks method is that we conduct Smote and Tomeklinks iteratively for T times (until reaching the criteria). Besides, to avoid excessive cleaning to the data in initial stages, we changed the rule of Tomeklinks in the first $T - 1$ times by removing the identified pairs under a certain probability (e.g., 0.35 in our work).

4   Methodology

We propose a spatial-temporal multi-view model for predicting crime occurrences, the overall process of which is shown in Fig. 1. The model consists of three main parts: Weather information generation, Statistical features, and Spatial relationship regularization.

4.1 Information Gathering

Online-Information. When extracting different kinds of features from tweets during the past time window (i.e., one week in this paper), we aim to exploit the latent relationships between theses features for crime prediction. Previous study [7,16] have shown the appealing ability of convolutional neural network in classification tasks. Therefore, we build a one-dimensional convolutional neural network for online information utilization. Our 1DCNN model consists of four layers: two convolutional layers followed by two pooling layers. The twitter features will be shared for all the crime types in a certain community.

Offline-Information. We build a Long short term memory based network (LSTM), which is suitable for processing time-serial data, to fully leverage historical crime observation. We feed a time window (which is set as one month) of historical crime data (a certain crime type e.g., theft) in a target community into an LSTM network.

The online information and offline information are exploited via 1DCNN and LSTM, respectively, and then unfolded into two one-dimensional vectors to be combined and fed into three fully connected layers, together with weather information for the final classification.

Weather Information. We additionally integrate weather information into our model to improve the performance. Specifically, we employ GANs [10] to generate the corresponding weather information. GANs can be formulated as a distribution-matching problem, which aims to minimize the distance between the learned weather distribution and an expected distribution. To relive GANs from the difficuly of handle time-series weather data we adopt TimeGAN [21]. The loss function of the TimeGAN $\mathcal{L}_G$ has two components $\mathcal{L}_S, \mathcal{L}_U$ as specified below:

$$\mathcal{L}_S = \mathbb{E}_{s,x_1:T \sim p}[\sum_t \|h_t - g_{\mathcal{X}}(h_S, h_{t-1}, z_t)\|_2]$$

$$\mathcal{L}_U = \mathbb{E}_{s,x_1:T \sim p}[\log y_S + \sum_t \log y_t] + \mathbb{E}_{s,x_1:T \sim \hat{p}}[\log(1 - \hat{y}_S) + \sum_t \log(1 - \hat{y}_t)]$$

where s, x_1 is the input following the distribution $p(S, X_{1:T})$, which is time-series data. $g_{\mathcal{X}}(h_S, h_{t-1}, z_t)$ is the output state of a LSTM network. y_S is the classification result from a softmax function. Hence, we define the final loss function for GAN as:

$$\mathcal{L}_G = \min_{\theta_g}(\eta \mathcal{L}_S + \max_{\theta_d} \mathcal{L}_U) \tag{1}$$

For temporal information, we use the binary cross-entropy as the loss function, as shown below:

$$\mathcal{L}_T = -\sum_{i=0} y_i \log(y_i') + \alpha |y_1 - W_G^K Y_1'| + \beta |y_1' - W_k^T Y_1'| \tag{2}$$

where W_i^G and W_i^T are the i-th row of geo-matrix and taxi flow matrix defined in the above section, Y_1' is a 77-dimensional vector in which each element Y_{1j}' denotes the predictive occurrence probability chosen from the softmax output of community j (Y_{1i}' is set to 0), α and β are the tuning weights for geographic and taxi flow influence, respectively. In particular, the second and third items can be regraded as a regularization for loss function to increase the robustness and to prevent overfitting; α and β will be tuned as hyper-parameters.

5   Experiments

We collect $152,460(90 \times 77 \times 22)$ samples of crime records spanning four months in Chicago. We set the time window to one month for historical crime data and one week for weather data. We train our model for each crime separately, as each crime has its own unique pattern and it does not improve the performance to consider training samples from other crime types. We split the dataset 90×77) of each crime type into training samples (5,005) and testing samples (1,935) by time. Therefore, testing samples contain the more recent data that training samples. Furthermore, we also conduct sampling and data cleaning for crime types, regarding which occurrence samples takes less then 20% of the total samples. Specifically, we apply iterative tomek-links to increase the percentage until it exceeds 20%. We grab historical weather information from a public source[4] and transform weather data into binary (indicating rain or not). We finally conduct experiments a machine with 8 GPUs and use accuracy as the evaluation metric.

5.1   Comparison Methods

- KDE(T): This model [9] combines features extracted from historical crime records using kernel density estimation (KDE) with topical features from twitter. Its experimental results shows improvement over 19 out of 26 crime types. We apply their model to our data using the same topical features used by our method to make a fair comparison. As we use natural community as region partition, our method does not require spatial interpolation as this method does.
- KDE(T*): This model [1] is a variant of the KDE(T). It models the temporal feature from twitter in a new way and combine them with historical crime records. We use the recommended settings in the original paper.
- KDE(T*+R): This model is an improved version of KDE(T*) and is also proposed by [1]. This model extracts routine activity patterns from twitter as additional features and feed them with KDE results into a binary classification method.
- TCP: TCP [22] is a framework that captures the intra-region temporal relationship and inter-region spatial relationship for crime prediction. We employ TCP on our features and transform its task from regression into classification. We set two parameters λ and ηH as 1 and 0.4, respectively, for our task.

---

[4] https://www.wunderground.com/.

- LSTM: We directly feed all features to this model for crime prediction, setting the hidden cells of LSTM layer to 70 and three fully-connected layers with 120, 60, 80 neurons following the LSTM layer.
- GRU: Gated Recurrent Unit (GRU) is a gating mechanism in recurrent neural networks [6]. It is similar to LSTM model but achieves better performance in many tasks. We feed all the features to this model and set hidden states of GRU layer to 70 and two fully-connected layers with 100, 50 neurons.

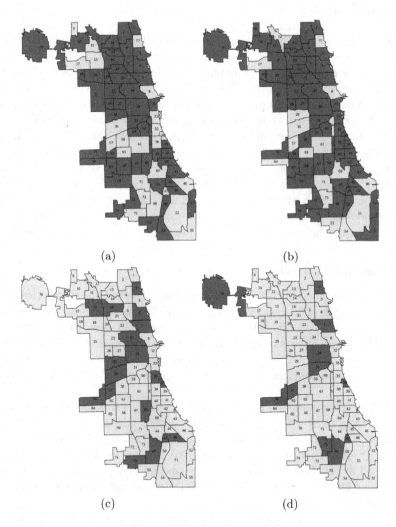

(a)   (b)

(c)   (d)

Fig. 2. True occurrences of theft over entire 12th (a); Predictive occurrences of theft over 12th (b); True occurrences of robbery over entire city on April 12th (c); Predictive occurrences of robbery over entire city on April 12th (d)

Table 2. Experimental results. CD denotes criminal damage; DP denotes deceptive practice; OO denotes other offenses; MVT denotes motor vehicle theft; CT denotes criminal trespass; WV denotes weapons violation; CO denotes children offense; CSA denotes crim sexual assault; PPV denotes public peace violation; PCI denotes public officer interference; LLV denotes liquor law violation; $\neg G$ denotes our model excluding geo-information; $\neg T$ denotes our model excluding taxi-information; $\neg W$ denotes our model excluding weather features.

Crime	SOTAs			DL based			Proposed			
	KDE(T*)	KDE(T*+R)	KDE(T)	TCP	GRU	LSTM	Ours	$\neg G$	$\neg T$	$\neg W$
Theft	0.66	0.69	0.68	0.70	0.68	0.66	**0.74**	0.722	0.729	0.728
Battery	0.72	0.75	0.74	0.73	0.76	0.76	**0.82**	0.777	0.792	0.810
CD	0.66	0.69	0.71	0.68	0.66	0.70	**0.75**	0.720	0.710	0.748
DP	0.68	0.72	0.67	0.71	0.64	0.66	**0.72**	0.677	0.676	0.711
Assault	0.66	0.72	0.73	0.71	0.69	0.70	**0.75**	0.743	0.733	0.741
OO	0.65	0.70	0.71	0.65	0.67	0.71	**0.77**	0.759	0.757	0.766
Narcotics	0.74	0.75	0.75	0.72	0.69	0.72	**0.77**	0.745	0.751	0.760
Burglary	0.67	0.66	0.67	0.71	0.65	0.69	**0.73**	0.701	0.677	0.709
MVT	0.62	0.66	0.67	0.66	0.63	0.62	**0.68**	0.667	0.662	0.681
Robbery	0.71	0.74	0.75	0.73	0.70	0.70	**0.78**	0.770	0.770	0.771
CT	0.66	**0.70**	0.69	**0.70**	0.67	0.67	0.66	0.646	0.621	0.651
WV	0.68	0.70	0.74	0.73	0.72	0.73	**0.77**	0.766	0.745	0.763
CO	0.59	0.62	0.60	0.63	0.60	0.61	**0.65**	0.641	0.631	0.641
CSA	0.66	0.67	0.65	0.64	0.62	0.63	**0.71**	0.684	0.681	0.702
PPV	0.60	0.63	0.65	0.64	0.61	0.62	**0.66**	0.655	0.658	0.643
PCI	0.69	0.72	**0.76**	0.74	0.70	0.71	**0.76**	0.732	0.743	0.752
SO	0.54	0.58	0.56	0.57	0.54	0.54	**0.59**	0.560	0.550	0.581
Prostitution	0.74	0.76	0.81	0.79	0.80	0.81	**0.87**	0.841	0.851	0.862
Homicide	0.61	0.59	0.62	0.63	0.61	0.63	**0.70**	0.674	0.662	0.692
Arson	0.53	0.57	0.57	0.54	0.51	0.51	**0.58**	0.544	0.533	0.571
LLV	0.63	0.62	**0.64**	0.59	0.56	0.56	0.63	0.581	0.575	0.620
Kidnapping	0.47	0.51	0.50	0.48	0.49	0.50	**0.53**	0.520	0.511	0.521

Our comparison results (Table 2) show the proposed method outperform the baseline methods on 19 out of 22 crime types.

5.2 Hyper-parameters Tuning

We set the hyperparameters of our model as follows: the number of hidden cells in LSTM layer as 30, filter number of convolutional layer as 10, filter length of filter of convolutional layer, number of neurons N1 as 60, number of neurons N2 as 80, number of neurons as 120, number of estimators in adaboosting as 5, learning rate of adaboosting as 0.7. For the GAN, we fix both the temporal correlation and feature correlations at 0.2.

5.3 Ablation Study

We conduct ablation study to investigate the effect of weather data, taxi information, and geo-information on our model's performance. The results are shown in the fourth column in Table 2.

6   Conclusion and Future Work

Crime prediction finds its root and rationale in the fact that many criminals tend to commit the same types of crimes proven successfully in similar time and locations. This paper presents a preliminary investigation of social and weather factors for crime activity prediction, and our experiments demonstrate the model's ability to improve the forecast of crime occurrences, along with a clear correlation between crime and weather forecasting. There are plenty of ways to improve the efficiency and accuracy of crime prediction using IoT sensors. In this regard, our future work will include obtaining hourly weather data instead of daily data from IoT sensors in small regions, using crime data that include time and postcode to match with sensor data, and adding non-weather sensor data (e.g., number of people passing, traffic at certain times, and noise level) to make more accurate predictions.

References

1. Al Boni, M., Gerber, M.S.: Predicting crime with routine activity patterns inferred from social media. In: 2016 IEEE International Conference on Systems, Man, and Cybernetics (SMC), pp. 001233–001238. IEEE (2016)
2. Bai, L., Yao, L., Kanhere, S.S., Wang, X., Liu, W., Yang, Z.: Spatio-temporal graph convolutional and recurrent networks for citywide passenger demand prediction. In: Proceedings of the 28th ACM International Conference on Information and Knowledge Management, pp. 2293–2296 (2019)
3. Batista, G.E., Prati, R.C., Monard, M.C.: A study of the behavior of several methods for balancing machine learning training data. ACM SIGKDD Explor. Newsl. **6**(1), 20–29 (2004)
4. Chawla, N.V., Bowyer, K.W., Hall, L.O., Kegelmeyer, W.P.: SMOTE: synthetic minority over-sampling technique. J. Artif. Intell. Res. **16**, 321–357 (2002)
5. Chen, X., Huang, C., Zhang, X., Wang, X., Liu, W., Yao, L.: Expert2Vec: distributed expert representation learning in question answering community. In: Li, J., Wang, S., Qin, S., Li, X., Wang, S. (eds.) ADMA 2019. LNCS (LNAI), vol. 11888, pp. 288–301. Springer, Cham (2019). https://doi.org/10.1007/978-3-030-35231-8_21
6. Cho, K., et al.: Learning phrase representations using rnn encoder-decoder for statistical machine translation. arXiv preprint arXiv:1406.1078 (2014)
7. Ding, X., Liu, T., Duan, J., Nie, J.Y.: Mining user consumption intention from social media using domain adaptive convolutional neural network. In: AAAI, vol. 15, pp. 2389–2395 (2015)
8. Fast, E., Chen, B., Bernstein, M.S.: Empath: understanding topic signals in large-scale text. In: Proceedings of the 2016 CHI Conference on Human Factors in Computing Systems, pp. 4647–4657 (2016)
9. Gerber, M.S.: Predicting crime using twitter and kernel density estimation. Decis. Support Syst. **61**, 115–125 (2014)
10. Goodfellow, I., et al.: Generative adversarial nets. In: Advances in Neural Information Processing Systems, pp. 2672–2680 (2014)
11. Gorman, D.M., Speer, P.W., Gruenewald, P.J., Labouvie, E.W.: Spatial dynamics of alcohol availability, neighborhood structure and violent crime. J. Stud. Alcohol. **62**(5), 628–636 (2001)

12. Kadar, C., Pletikosa, I.: Mining large-scale human mobility data for long-term crime prediction. EPJ Data Sci. **7**(1), 1–27 (2018). https://doi.org/10.1140/epjds/s13688-018-0150-z

13. Kenrick, D.T., MacFarlane, S.W.: Ambient temperature and horn honking: a field study of the heat/aggression relationship. Environ. Behav. **18**(2), 179–191 (1986)

14. Li, Y., Liu, Z., Yao, L., He, Z.: Non-local self-attentive autoencoder for genetic functionality prediction. In: Proceedings of the 29th ACM International Conference on Information & Knowledge Management, pp. 2117–2120 (2020)

15. Ning, X., Yao, L., Benatallah, B., Zhang, Y., Sheng, Q.Z., Kanhere, S.S.: Source-aware crisis-relevant tweet identification and key information summarization. ACM Trans. Internet Technol. (TOIT) **19**(3), 1–20 (2019)

16. Ning, X., Yao, L., Wang, X., Benatallah, B.: Calling for response: automatically distinguishing situation-aware tweets during crises. In: Cong, G., Peng, W.-C., Zhang, W.E., Li, C., Sun, A. (eds.) ADMA 2017. LNCS (LNAI), vol. 10604, pp. 195–208. Springer, Cham (2017). https://doi.org/10.1007/978-3-319-69179-4_14

17. Ning, X., Yao, L., Wang, X., Benatallah, B., Salim, F., Haghighi, P.D.: Predicting citywide passenger demand via reinforcement learning from spatio-temporal dynamics. In: Proceedings of the 15th EAI International Conference on Mobile and Ubiquitous Systems: Computing, Networking and Services, pp. 19–28 (2018)

18. Tomek, I., et al.: An experiment with the edited nearest-nieghbor rule (1976)

19. Wang, H., Kifer, D., Graif, C., Li, Z.: Crime rate inference with big data. In: Proceedings of the 22nd ACM SIGKDD International Conference on Knowledge Discovery and Data Mining, pp. 635–644 (2016)

20. Yan, X., Guo, J., Lan, Y., Cheng, X.: A biterm topic model for short texts. In: Proceedings of the 22nd International Conference on World Wide Web, pp. 1445–1456 (2013)

21. Yoon, J., Jarrett, D., van der Schaar, M.: Time-series generative adversarial networks. In: Advances in Neural Information Processing Systems, pp. 5508–5518 (2019)

22. Zhao, X., Tang, J.: Modeling temporal-spatial correlations for crime prediction. In: Proceedings of the 2017 ACM on Conference on Information and Knowledge Management, pp. 497–506 (2017)

Experimental Analysis of Locality Sensitive Hashing Techniques for High-Dimensional Approximate Nearest Neighbor Searches

Omid Jafari$^{(\boxtimes)}$ ⓘ and Parth Nagarkar ⓘ

New Mexico State University, Las Cruces, USA
{ojafari,nagarkar}@nmsu.edu

Abstract. Finding nearest neighbors in high-dimensional spaces is a fundamental operation in many multimedia retrieval applications. Exact tree-based approaches are known to suffer from the notorious *curse of dimensionality* for high-dimensional data. Approximate searching techniques sacrifice some accuracy while returning *good enough* results for faster performance. Locality Sensitive Hashing (LSH) is a popular technique for finding approximate nearest neighbors. There are two main benefits of LSH techniques: they provide theoretical guarantees on the query results, and they are highly scalable. The most dominant costs for existing external memory-based LSH techniques are algorithm time and index I/Os required to find candidate points. Existing works do not compare both of these costs in their evaluation. In this experimental survey paper, we show the impact of both these costs on the overall performance. We compare three state-of-the-art techniques on six real-world datasets, and show the importance of comparing these costs to achieve a more fair comparison.

Keywords: Locality Sensitive Hashing · High-dimensional spaces · Approximate nearest neighbor

1 Introduction

Many large multimedia retrieval applications require efficient processing of nearest neighbor queries in high-dimensional spaces. Exact tree-based indexing structures, such as KD-tree, SR-tree, etc., work well for low-dimensional spaces (<10) but suffer from the notorious *curse of dimensionality* for high-dimensional spaces. They are often outperformed by brute-force linear scans [4]. One solution to this problem is to search for *good enough* approximate results instead. Approximate techniques sacrifice some accuracy for a significant improvement in the overall processing time. In many applications where 100% is not needed, this tradeoff is very useful in saving time. The goal of the approximate version of the nearest neighbor problem, also called *c-approximate Nearest Neighbor search*, is

© Springer Nature Switzerland AG 2021
M. Qiao et al. (Eds.): ADC 2021, LNCS 12610, pp. 62–73, 2021.
https://doi.org/10.1007/978-3-030-69377-0_6

to return points that are within $c * R$ distance from the query point. Here, $c > 1$ is a user-defined approximation ratio and R denotes the distance of the query point and its nearest neighbor.

1.1 Locality Sensitive Hashing

Locality Sensitive Hashing (LSH) [8] is one of the most popular techniques for finding approximate nearest neighbors in high-dimensional spaces. LSH was first introduced in [8] for the Hamming distance, but was later extended to several distances, such as the popular Euclidean distance [6]. LSH uses *random* hash projections to map the original high-dimensional space to the projected low-dimensional space. The main idea behind LSH is that nearby points in the original high-dimensional space will map to similar hash buckets in the low-dimensional space with a higher probability than dissimilar or far away points. Since LSH was first proposed in [8], there have been several works that have focused on improving the search accuracy and/or performance [3,7,9,15,16,18,23].

1.2 Motivation for Using LSH

Locality Sensitive Hashing (LSH) is known for two main advantages: its sublinear query performance (in terms of the data size) and theoretical guarantees on the query accuracy. Additionally, LSH uses random hash functions which are data-independent (i.e. data properties such as data distribution are not needed to generate these random hash functions), and the generation of these hash functions is a simple process that takes negligible time. Additionally, the data distribution does not affect the generation of these hash functions. Hence, in applications where data is changing or where newer data is coming in, these hash functions do not require any change during runtime. While the original LSH index structure suffered from large index sizes (in order to obtain a high query accuracy) [3,18], state-of-the-art LSH techniques [7,9] have alleviated this issue by using advanced methods such as *Collision Counting* and *Virtual Rehashing*. In addition to their fast index maintenance, fast query performance, and theoretical guarantees on the query accuracy, LSH algorithms are easy to implement as external memory-based algorithms, and hence are more scalable than in-memory algorithms (such as graph-based ANN algorithms) [15].

1.3 Motivation of Our Experimental Survey

Locality Sensitive Hashing techniques have two dominant costs for finding nearest neighbors: 1) cost of reading the index files from the external memory to the main memory (which we call *Index I/Os*), and 2) cost of finding candidates and removing false positives (which we call *Algorithm time*). As mentioned in Sect. 1.2, one of the benefits of LSH is that it is a scalable algorithm. Some of the existing LSH techniques (e.g. C2LSH [7] and QALSH [9]) are not entirely external memory-based (i.e. even though the indexes are stored on the disk, their implementations require the entire data and indexes should fit into the main memory during the index creation phase). Thus, existing works (such as [1]) do not compare their results with C2LSH and QALSH on large datasets since they do not fit in the main memory. Additionally, some recent works (such as [15]) only compare the *Index I/Os* without comparing the important *Algorithm time*. This leads to other recent papers (such as [13,14,25]) to unfairly compare their *Algorithm time* with QALSH or I-LSH [15] since they are deemed as the state-of-the-art LSH techniques.

1.4   Contributions of This Experimental Survey Paper

In this paper, we carefully present a detailed experimental analysis on three state-of-the-art *external memory-based* LSH algorithms, C2LSH [7], QALSH [9], and I-LSH [15]. Our contributions are as follows:

- We modify the implementations of C2LSH and QALSH to create fully external memory-based implementations such that the entire dataset and/or the entire index do not need to be in the main memory for the algorithms to work during index generation or query processing.[1]
- We show the importance of experimentally analyzing and comparing the *Index I/Os* and *Algorithm time* of all algorithms.
- We compare these three algorithms on real datasets with different characteristics under differing system parameters.

To the best of our knowledge, we are the first work to present a detailed analysis of these three state-of-the-art LSH techniques.

2   Related Work

Nearest Neighbor problem is an important problem for multimedia applications in many diverse domains such as multimedia retrieval, image processing, machine learning, etc. Since tree-based index structures can be outperformed by a linear scan, due to the *curse of dimensionality*, in high-dimensional spaces, approximate techniques are preferred due to their fast performance at the expense of some accuracy. These techniques can be broadly classified into three main categories: Hashing-based methods, Partition-based methods, and Graph-based methods.[2] Hashing-based methods can be further classified into learning-based hashing techniques and random hashing techniques. The benefit of random hashing techniques, such as Locality Sensitive Hashing [8], are that they are easy to construct, no need for training data, and easy to maintain and update. Additionally, LSH provides a sub-linear (in terms of the data size) query performance and theoretical guarantees on the query accuracy.

---

[1] These implementations will be made public.
[2] We refer the reader to a recent survey [14] for an in-depth survey on these categories.

Locality Sensitive Hashing and its Variants: The main idea of Locality Sensitive Hashing is to create random projections and hash data points in these random projections such that nearby data points in the original high-dimensional space will be mapped to the same hash bucket with a higher probability compared to data points that are far apart from each other. It was originally proposed in [8] for the Hamming distance and then later extended to the popular Euclidean distance [6]. In this original work on Euclidean distance (E2LSH), instead of a single hash function (or a projection), a hash table consisted of several hash functions (represented by Compound Hash Keys) was built to reduce false positives. But this also generated false negatives. Hence several hash tables had to be used to reduce the number of false positives and false negatives, while keeping the accuracy of the query high. The main drawbacks of this approach were the size of the index structure (since large number of hash tables were required to return the desired number of results with a high accuracy) and the need to determine the width of the hash bucket during index creation (a larger width returned enough results but also with a potential of too many false positives, whereas a smaller width had a potential of misses resulting in insufficient results). This user-defined width, which was mainly dependent on the data distribution, had to be often determined through a trial and error process. LSH-Forest [3] was proposed where the compound hash-keys were hierarchically stored such that the algorithm could stop at a higher level in the tree if more results were needed. In Multi-probe LSH [18], the authors proposed a technique to probe into neighboring buckets when more results were needed. The intuition is that neighboring buckets are more likely to contain nearby points. Hence, if the bucket width was underestimated (which is better than overestimation which can lead to significant wasteful processing), neighboring buckets were probed to find the desired number of results.

Later, C2LSH [7] introduced two main concepts of *Collision Counting* and *Virtual Rehashing* that solved the two main drawbacks of E2LSH [6]. In C2LSH, the authors proposed to create m base hash functions and choose candidate points based on how many times a data point collides with the query point (and hence instead of creating several hash tables of several hash functions, only 1 table of m base hash functions is needed), which reduced the size of the index structure. Additionally, in *Virtual Rehashing*, the neighboring buckets in each hash function are read incrementally when sufficient number of results are not found. In SK-LSH [16], the authors propose a linear ordering on the Compound Hash Keys (using a space-filling curve) such that nearby Compound Hash Keys are stored on the same (or nearby) page on the disk, thus reducing the total number of I/Os. The design of SK-LSH is still build on the original E2LSH, and hence suffers from the parameter tuning problem, where the user is expected to enter important parameters such as number of hash functions and the radius at which k results will be found. QALSH [9] was later proposed that built query-aware hash functions such that the hash value of the query point is considered as the anchor bucket during query processing and this idea would solve the issue when close points to a query were partitioned into different buckets when query was near the bucket boundaries. Additionally, B+trees are built on each hash function for efficient lookups into neighboring buckets (which translate to range queries). QALSH utilizes the concepts of *Collision Counting* and *Virtual Rehashing*. HD-Index [1] was introduced which generated Hilbert keys of the dataset points and also stored the distances of points to each other to efficiently prune the results based on distance filters. Due to the reliance on space-filling curves (Hilbert curves) and B+-trees, HD-Index cannot scale for moderately high-dimensional datasets [1]. SRS [22] uses the Euclidean distance between two points in the projected space to estimate their distance in the original space. In order to find the next nearest neighbor in the projected space, SRS uses an R-tree to index the points in the projected space. This incremental finding of the NN is similar to I-LSH. The main goal of SRS is to introduce a very lightweight index structure to solve the ANN problem. SRS is shown to suffer from memory leaks and slow running times as compared with C2LSH [1], and hence not included in our work. Recently, I-LSH [15], which is considered to be the state-of-the-art LSH technique [13], was proposed to improve the Virtual Rehashing process of QALSH (where the range of the lookups are incremented exponentially). In I-LSH, the authors propose to increase the range of the lookups based on the distance to the nearest point (in the projected space) instead of increasing the range exponentially. While this strategy results in less disk I/Os, it also leads to high disk seeks (random I/Os) and algorithm time as we show in Sect. 4. Very recently, an in-memory LSH algorithm, PM-LSH [25] was proposed where the idea was to estimate the Euclidean distance based on a tunable confidence interval value such that the overall query processing time is reduced.

3   State-of-the-Art Techniques

In this section, we will introduce the concepts introduced by the three state-of-the-art external memory-based LSH techniques, C2LSH [7], QALSH [9], and I-LSH [15]. We primarily use the terminologies and formulations introduced in E2LSH [6] and C2LSH [7]. Due to space limitations, we ask the reader to refer to [7] for detailed formulations. C2LSH [7] introduced the concepts of *Collision Counting* and *Virtual Rehashing*. In [7], authors theoretically show that two close points x and y collide in at least l hash layers (out of m hash layers) with a probability $1 - \delta$. Further, only those points that collide at least l times with the query point, where l is the collision count threshold, are chosen as candidates. C2LSH creates only one hash function per hash table, and hence the number of hash functions are equal to the number of hash table.

Instead of assuming a *magic* radius (which traditional LSH methods did), C2LSH sets the initial radius R to 1. It is possible that with $R = 1$, there are not enough results for a top-k query to be returned. C2LSH increases the radius of the query in the following sequence: $R = 1, c, c^2, c^3....$ If at *level-R*, enough candidates are not found, the radius is increased until enough query results are found. This exponential expansion process is called *Virtual Rehashing*.

Moreover, C2LSH uses two terminating conditions to stop the algorithm. These conditions specify that 1) at the end of each virtual rehashing at least k

candidates should have been found whose Euclidean distance to the query are less than or equal to cR, and 2) at any point, $k + \beta n$ candidates are found.

QALSH introduces *query-aware* hash functions. For a query q, once the query projection is found by computing $h_a(q)$, QALSH uses the query as the "anchor" to find the anchor bucket with width w with the interval $|h_a(q) - \frac{w}{2}, h_a(q) + \frac{w}{2}|$. If the projected location for a point x falls in the same anchor bucket as q, i.e., $|h_a(o) - h_a(q)| \leq \frac{w}{2}$, then QALSH considers that o has collided with q under h_a. QALSH [9] also utilizes these concepts of Collision Counting and Virtual Rehashing to build *query-aware* hash functions. Another main difference of QALSH is that it uses B+-trees to represent the hash tables. An exponential expansion in each hash table is thus the same as a range query on a B+-tree. By using *query-aware* hash functions and B+-trees, QALSH improves the theoretical bounds by reducing the total number of hash functions required to satisfy the quality guarantee. Additionally, QALSH can work for any approximation ratio, c, greater than 1, while C2LSH can only work for $c \geq 2$. While the reduction in number of hash functions generates a smaller index, the overhead of using B+-trees makes QALSH much slower as we experimentally show in Sect. 4.

I-LSH [15] uses query-aware hash functions (proposed by QALSH) and proposes an incremental expansion strategy to reduce overall index I/Os. In order to do that, I-LSH finds the next closest point in each projection. While this process leads to less overall index I/Os, it still requires disk seeks and (as we show in Sect. 4) the algorithm overhead is far more than the savings in the disk I/Os.

4   Experimental Analysis

In this section, we first explain our experimental evaluation plan. We experimentally analyze C2LSH, QALSH, and I-LSH on different datasets and report the results for varying criteria. All experiments were run on the nodes of the Bigdat cluster[3] with the following specifications: two Intel Xeon E5-2695, 256 GB RAM, and CentOS 6.5 operating system. All codes were written in C++11 and compiled with gcc v4.7.2 with the - O3 optimization flag. As mentioned in Sect. 1.4, we extend the implementations of C2LSH and QALSH to be completely external-memory based implementations (i.e. the entire dataset or the index files are not needed to be in the main memory in order to construct the LSH indexes).

4.1   Datasets

We use the following six diverse high-dimensional datasets in our experiments:

- **P53** [5] consists of $31,002$ 5409-dimensional points which are generated based on the biophysical features of mutant p53 proteins.
- **LabelMe** [19] consists of $181,093$ 512-dimensional points which were generated by running the GIST feature extraction algorithm on annotated images.

---

[3] Supported by NSF Award #1337884.

- **Sift1M** [10] consists of $1,000,000$ 128-dimensional points that were created by running the SIFT feature extraction algorithm on real images.
- **Deep1M** consists of $1,000,000$ 96-dimensional points sampled from the Deep1B dataset introduced in [2].
- **Mnist8M** [17] This dataset contains $8,100,000$ 784-dimensional points that represent images of the digits 0 to 9 which are grayscale and of size 28×28.
- **Tiny80M** [24] This dataset contains $79,302,017$ 384-dimensional points generated using Gist feature extraction algorithm on 80 million colored images.

4.2   Evaluation Criteria and Parameters

The goal of our paper is to present a detailed analysis of the performance and accuracy of the state-of-the-art LSH techniques. We randomly choose 50 queries and report the average of the results. We used the same parameters suggested in their papers ($w = 2.781$ for QALSH and $w = 2.184$ for C2LSH). We choose $\delta = 0.1$ and $c = 2$ (since C2LSH cannot give guarantees for $c < 2$). Since I-LSH uses the same hash functions as QALSH, their index size and index construction time are the same. [9] shows the difference between these two criteria for C2LSH and QALSH for different datasets, and hence we avoid it in this paper.

After careful analysis of performance of LSH techniques, we present the following breakdown of the query processing time (QPT):

- **Index Read Cost:** LSH techniques need to read index files (from the external memory) in order to find the candidates. This dominant cost of reading index files can be further broken down into the number of disk seeks (i.e. random I/Os) and the total amount of data read. Following [15], we also consider the number of disk seeks and amount read in our cost formulation.
- **Algorithm Time:** Another dominant cost in LSH processing is the processing of index files once they are read into the main memory. LSH techniques need to find points that are considered as candidates. Techniques such as Collision Counting (explained in Sect. 3) are included in this cost.
- **False Positive Removal Cost:** Once a point is deemed as a candidate, its Euclidean distance with the query point is calculated. Since the state-of-the-art LSH techniques have an upper bound of the number of candidates (which is set to $k + 100$), this cost is negligible as compared to the previous two costs. Due to space limitations and since we observed that this cost is less than 0.5 ms for all algorithms, we do not show the results of this cost.

It is well-known that random I/Os are much more expensive than sequential I/Os [12]. Additionally, the difference in the cost changes significantly depending on whether the external storage medium is an HDD or an SSD. The difference in the costs of random I/Os and sequential I/Os is significantly more in HDDs than in SSDs (mainly because random disk seeks are faster in SSDs than HDDs) [11]. We noticed that the number of disk seeks are significantly different in these LSH techniques due to how they find neighboring points in projected spaces. Hence, we model the overall *Query Processing Time* (QPT) for both HDDs and SSDs.

For an HDD, we use the reported benchmarks for Seagate Barracuda HDD with 7200 RPM and 1TB: average disk seek requires 8.5 ms and the average data read rate is 0.156 MB/ms [21]. Similarly, for an SSD, we use the reported benchmarks for the Seagate Barracuda 120 SSD with 1TB storage: average disk seek requires 0.01 ms and the average data read rate is 0.56 MB/ms [20].

We use the same accuracy measure, the overall ratio, used in several prior works [7,9,15,16]: $\frac{1}{k}\sum_{i=1}^{k}\frac{||o_i,q||}{||o_i^*,q||}$. Here, o_i is the ith point returned by the technique and o_i^* is the true ith nearest point from q (ground truth). The closer the ratio is to 1, the higher is the accuracy of the LSH technique.

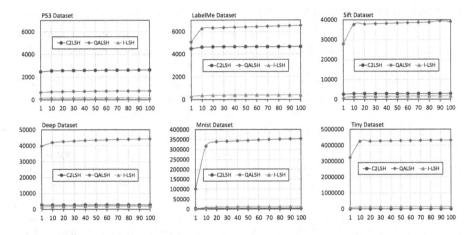

Fig. 1. Number of disk seeks (Y axis) for different k (X Axis) on 6 datasets

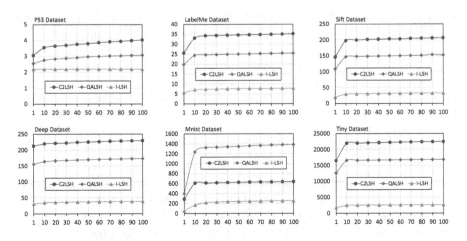

Fig. 2. Amount of data read (in MB) (Y axis) for k (X Axis) on 6 datasets

4.3 Discussion of the Performance Results

Number of Disk Seeks: Figure 1 shows the required number of disk seeks (random I/Os). We observed that the performance of I-LSH degrades as dataset size becomes large. This is because I-LSH needs to find the closest projected point each time the radius needs to be expanded, which further requires reading the indexed points from the disk several times. We also observe that QALSH has a better performance compared to C2LSH for smaller datasets, but as the dataset size increases, the number of seeks are significantly higher than C2LSH and I-LSH. This is happening because the search radiuses of QALSH are larger than C2LSH in larger datasets, which results in higher disk seeks.

Amount of Data Read: Figure 2 shows the total amount of data that was read from the index files. I-LSH always has the least amount of data read for all datasets because it incrementally searches for the nearest points in the projections instead of having buckets and fixed widths. However, we later show that these I/O savings are offset by the processing time of finding these nearest points. C2LSH reads more data than QALSH for most datasets (except Mnist) since QALSH uses less hash projections because they are query-aware.

Algorithm Time: Figure 3 shows the time needed to find the candidates (excluding the I/O times). This figure shows the huge overhead of I-LSH which is caused due to their incremental searching strategy. Also, since I-LSH and QALSH both use B+-trees, which become huge for the larger datasets, their performance degrades heavily in these cases. Since C2LSH does not have any overhead of additional index structures (such as B+-tree), it has the least Algorithm time for all datasets. In terms of Algorithm Time, I-LSH is faster than QALSH (except for the P53 dataset - which is the smallest dataset in our experiments) mainly because it has to process less hash functions than QALSH [15].

Query Processing Time (on HDD): Figure 4 shows the overall time required to solve a given k-NN query on a Hard Disk Drive. I-LSH performs the best for smaller datasets (P53 and LabelMe) because its Algorithm Time overhead is small, but as the dataset size increases, the Algorithm Time overhead offsets the savings in disk seeks and performs worse than C2LSH (but better than QALSH). Except for the smallest dataset (P53), QALSH is the slowest of the three algorithms. It works good for smaller datasets (P53) but does not scale well for moderate and large sized datasets. For larger datasets, C2LSH is always the fastest technique since its having better algorithm time and number of disk seeks compared to the other two algorithms.

Query Processing Time (on SSD): Figure 5 shows the overall time required to solve a given k-NN query on a Solid State Drive. In SSDs, I/O operations are much faster and the overall Query Processing Time is mainly dominated by the algorithm time. Therefore, C2LSH (which has the best Algorithm time) always performs the best on SSDs (for all datasets) followed by I-LSH (except for the smallest dataset, P53).

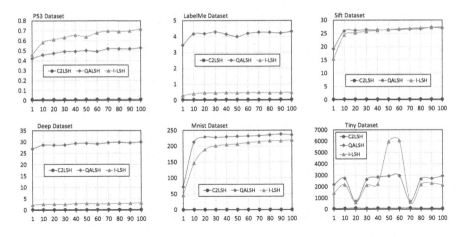

Fig. 3. Algorithm time (in s) (Y axis) for k (X Axis) on 6 datasets

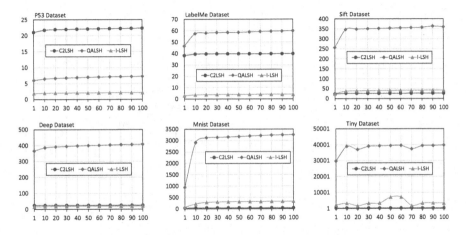

Fig. 4. HDD query processing time (in s) (Y axis) for k (X Axis) on 6 datasets

Accuracy Ratio: Figure 6 shows the accuracy of the compared techniques. Ratios are always greater than or equal to 1 and having a ratio equal to 1 equates to the highest accuracy. Except for the Mnist dataset, C2LSH produces the best accuracy among the three algorithms. QALSH is more accurate than I-LSH, which we believe is mainly because it uses more hash functions than I-LSH. Except for C2LSH's accuracy on the Mnist dataset, all three algorithms produce accurate results for all datasets.

Overall, we find that C2LSH can find k-NN results faster than QALSH and I-LSH, mainly because of the simplicity of their hash functions (i.e. an additional index structure, B+-tree, is not used). Additionally, all three algorithms produce accurate results (with C2LSH producing slightly better accurate results than QALSH and I-LSH for most datasets).

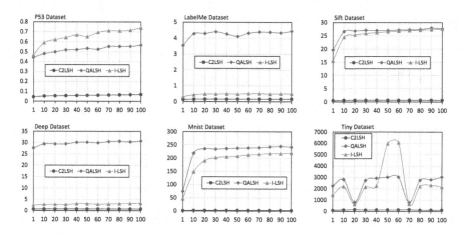

Fig. 5. SSD query processing time (in s) (Y axis) for k (X Axis) on 6 datasets

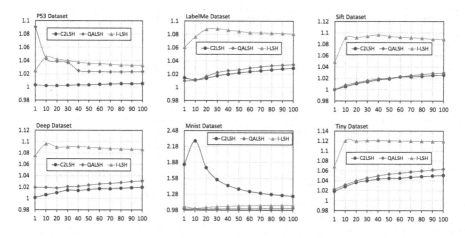

Fig. 6. Accuracy ratio (Y axis) for different k (X Axis) on 6 datasets

5   Conclusion

Locality Sensitive Hashing is a popular technique for solving Approximate Nearest Neighbor queries in high-dimensional spaces. In this paper, we presented a detailed experimental analysis on three popular state-of-the-art LSH algorithms, C2LSH, QALSH, and I-LSH. We presented our analysis on diverse datasets with varying characteristics (cardinality and dimensionality). We show that while reducing disk seeks is important, it cannot be at the expense of Algorithm time, which can be a dominant cost in the overall query processing time for large datasets. We importantly show that improvements in one aspect of the LSH workflow (e.g. disk seeks), does not necessarily result in overall query processing performance improvement.

References

1. Arora, A., et al.: HD-index: pushing the scalability-accuracy boundary for approximate knn search. In: VLDB (2018)
2. Babenko, A., et al.: Efficient indexing of billion-scale datasets of deep descriptors. In: CVPR (2016)
3. Bawa, M., et al.: LSH forest: self-tuning indexes for similarity search. In: WWW (2005)
4. Chávez, E., et al.: Searching in metric spaces. CSUR **33**, 273–321 (2001)
5. Danziger, S.A., et al.: Predicting positive P53 cancer rescue regions using most informative positive (MIP) active learning. PLoS Comput. Biol. **5**, e1000498 (2009)
6. Datar, M., et al.: Locality-sensitive hashing scheme based on p-stable distributions. In: SOCG (2004)
7. Gan, J., et al.: Locality-sensitive hashing scheme based on dynamic collision counting. In: SIGMOD (2012)
8. Gionis, A., et al.: Similarity search in high dimensions via hashing. In: VLDB (1999)
9. Huang, Q., et al.: Query-aware locality-sensitive hashing for approximate nearest neighbor search. VLDB **9**, 1–12 (2015)
10. Jegou, H., et al.: Product quantization for nearest neighbor search. TPAMI **33**, 117–128 (2010)
11. Kim, A., et al.: Optimally leveraging density and locality for exploratory browsing and sampling. In: HILDA (2018)
12. Leis, V., et al.: Query optimization through the looking glass, and what we found running the join order benchmark. VLDB **27**, 643–668 (2018)
13. Li, M., et al.: I/o efficient approximate nearest neighbour search based on learned functions. In: ICDE (2020)
14. Li, W., et al.: Approximate nearest neighbor search on high dimensional data - experiments, analyses, and improvement. TKDE (2019)
15. Liu, W., et al.: I-LSH: I/O efficient c-approximate nearest neighbor search in high-dimensional space. In: ICDE (2019)
16. Liu, Y., et al.: SK-LSH: an efficient index structure for approximate nearest neighbor search. VLDB **7**, 745–756 (2014)
17. Loosli, G., et al.: Training invariant support vector machines using selective sampling. Large Scale Kernel Mach. (2007)
18. Lv, Q., et al.: Multi-probe LSH: efficient indexing for high-dimensional similarity search. In: VLDB (2007)
19. Russell, B.C., et al.: LabelMe: a database and web-based tool for image annotation. IJCV **77**, 157–173 (2008)
20. Seagate Barracuda 120 SSD Manual. https://www.seagate.com/www-content/datasheets/pdfs/barracuda-120-sata-DS2022-1-1909US-en_US.pdf
21. Seagate ST2000DM001 Manual. https://www.seagate.com/files/staticfiles/docs/pdf/datasheet/disc/barracuda-ds1737-1-1111us.pdf
22. Sun, Y., et al.: SRS: solving c-approximate nearest neighbor queries in high dimensional euclidean space with a tiny index. VLDB (2014)
23. Tao, Y., et al.: Efficient and accurate nearest neighbor and closest pair search in high-dimensional space. TODS **35**, 1–46 (2010)
24. Torralba, A., et al.: 80 million tiny images: a large data set for nonparametric object and scene recognition. TPAMI **30**, 1958–1970 (2008)
25. Zheng, B., et al.: PM-LSH: a fast and accurate LSH framework for high-dimensional approximate NN search. VLDB **13**, 643–655 (2020)

ANSWER: Generating Information Dissemination Network on Campus

Qing Qing[1], Teng Guo[1], Dongyu Zhang[1], and Feng Xia[2(✉)]

[1] School of Software, Dalian University of Technology, Dalian 116620, China
[2] School of Engineering, IT and Physical Sciences,
Federation University Australia, Ballarat, VIC 3353, Australia
f.xia@ieee.org

Abstract. Information dissemination matters, both on an individual and group level. For college students who are physically and mentally immature, they are more sensitive and susceptible to unnormal information like rumors. However, current researches focus on large-scale online message sharing networks like Facebook and Twitter, rather than profile the information dissemination on campus, which fail to provide any references for daily campus management. Against this background, we propose a framework to generate the information dissemination network on campus, named ANSWER (cAmpus iNformation diSsemination netWork gEneRation), based on multimodal data including behavior data, appearance data, and psychological data. The construction of the ANSWER is listed as four steps. First, we use a convolutional autoencoder to extract the students' facial features. Second, we process the behavior data to construct a friendship network. Third, heterogeneous information is embedded in the low-dimensional vector space by using network representation learning to obtain embedding vectors. Fourth, we use the deep learning model to predict. The experiment results show that ANSWER outperforms other methods in multiple feature fusion and prediction of information dissemination relationship performance.

Keywords: Information dissemination · Attribute network · Network generation

1 Introduction

Information dissemination matters, both on an individual and group level. The sharing of good news can spread happiness [12], while the diffusion of bad news, like rumors, could cause serious consequences [27]. For example, during this pandemic, the spread of misinformation about COVID-19 is impeding healthy behaviors and promoting erroneous practices that facilitate the spread of the virus and result in poor physical and mental health situation among individuals [25]. College students are a very special group of people in their emerging adulthood. In this stage, teenagers pursue personality exploration and may play

M. Qiao et al. (Eds.): ADC 2021, LNCS 12610, pp. 74–86, 2021.
https://doi.org/10.1007/978-3-030-69377-0_7

many roles such as college students, full-time employees. Commonly, they are leaving their families and toward independence [21]. Thus, various information will affect their physical and mental health deeply. In this case, simulating information dissemination patterns on campus and exploring the mechanisms behind them is an urgent research topic for education-related research fields.

However, exploring the hidden modes of information dissemination among students is a complex issue. Serving as the traditional method of collecting data for information dissemination, the amount of data collected by the questionnaire is small and time-consuming. At present, the researches on massive data are mainly based on online platforms such as Twitter [14], Facebook [12]. Due to privacy and security issues, researchers in the school administration fail to access these data, and cannot regard these data as a reference for academic research [9,13]. Moreover, the information itself is so private that the previous research is not enough to analyze the information dissemination between students.

Different from traditional data mining, multimodal data presents more complex information [4,15]. Furthermore, the advancement of network science has provided us with great help in analyzing social relationships, enabling us to abstract real-world relationships into the structure of the network [10]. These two major advancements provide us with an opportunity to analyze the hidden relationships behind students' information dissemination. Nevertheless, new challenges have emerged. The choices they make when choosing friends to disseminate information affected by many factors [29], thus need to select the appropriate approach. Moreover, the underlying network structure of information dissemination between students is invisible, and there must be an interaction between two people, which can be inferred from the information flow. But it is difficult to obtain disseminated data among students because of its privacy. Yet information dissemination is closely related to the common friendship network. People usually tell friends when they know information, but not each friend. Therefore, the prediction of information dissemination relationships can be developed by using the friendship network.

In this paper, we focus on students' friendship networks and combine students' facial and psychological attributes to construct a friend attribute network to complete the prediction of campus information dissemination relationships. The model named ANSWER (cAmpus iNformation diSsemination netWork gEneRation) is divided into four parts, as shown in Fig. 1. First, through the co-occurrence processing of the consumption record data of the student's campus card, the adjacency matrix of the friendship network is constructed. Secondly, we use the convolutional autoencoder to extract features from student facial images and express them as dense vectors. Thirdly, we use Network Representation Learning to embed the adjacency matrix of the friend attribute network. The network structure and node attributes of two heterogeneous information sources are processed in the same vector space. Fourth, a three-layer neural network is used to predict the information dissemination relationships.

In summary, our contributions can be summarized as follows:

(1) Based on the spatial and temporal behavioral data recorded by the campus card, we build a friendship network adjacency matrix containing only 0 or 1 by using the co-occurrence processing.
(2) We use the convolutional autoencoder to process the facial image data, to get the vector representation of facial features.
(3) We use facial features and psychographic data as the node attributes of the friendship network. The experimental results highlight the outstanding capabilities of ANSWER for information dissemination prediction.
(4) ANSWER is designed to solve the problem of information dissemination in the campus environment, mainly utilizes the friend attribute network to complete the construction of the information dissemination network. It combines the network topology and node attributes, and its performance is significantly better than other approaches.

This paper is organized as follows. In the next Section, we discuss the latest researches developments in relevant theoretical work. In Sect. 3, we describe the details of the problem formulation. In Sect. 4, we present the methodology in the proposed framework. In Sect. 5, we introduce the dataset and the analysis of the experimental results. We conclude the paper in Sect. 6.

2   Related Work

2.1   Network Representation Learning (NRL)

Network Representation Learning (also known as graph embedding) is based on a mapping function that is dedicated to solving the problem of data sparsity. The mapping function converts nodes into low-dimensional vectors to intuitively represent the relationship between the nodes in the original graph and finally used for downstream network analysis tasks. Perozzi et al. [22] carried out the first method of embedding, namely DeepWalk based on network structure. Its core draws on the Word2vec and the collinear relationship between nodes, uses a random walk strategy to generate node sentences based on the weight of edges, and uses Skip-gram training to obtain word vectors. Based on the Deepwalk theory, many researchers have made improvements [24,32].

These approaches are all based on the network structure to obtain the vector representation of nodes. However, there are also attribute features. Hou et al. [8] conducted a graph representation learning framework called Property Graph Embedding (PGE), which considers the relationship between the network structure and the attribute features.

2.2   Social Relationship

In social sciences, relationships are divided into two types: one-way (celebrities and ordinary people) and two-way (friends) [7]. Social structure is mainly connected by social relationships. By analyzing the formation in social relationships,

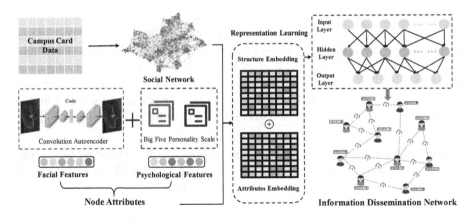

Fig. 1. Illustration of prediction framework ANSWER.

we can better understand the dynamic changes in the social structure. Previous researches on social relationships can be divided into two categories, link prediction [20] and relation type prediction [11,16]. Link prediction occupies an important role in social networks, which can be used to reconstruct complex networks. The types of social relationships are segmented into two types according to their nature, one is the blood relationship, etc., the other is the relationship formed by many social behaviors such as academic collaboration, information dissemination. Zhao et al. [31] identified vulnerable relationships in the information dissemination network, and showed that vulnerable relationships can largely affect information sharing, but have little impact on the information exchange network.

2.3 Information Dissemination Network

Existing literature exploring information dissemination focuses on online social platforms such as Twitter [14], Facebook [12]. Lerman et al. [14] constructed the social network for active users on Digg and Twitter, and analyzed the patterns of information dissemination. Elisa et al. [19] reconstructed the information dissemination network using natural language processing to study the dynamics of information dissemination and found that the dynamics were similar to epidemic SIR. Campan et al. [2] studied the dissemination of fake news in online social networks and analyzed its dissemination and influence. The commonality of the above-mentioned literature researches is that the network construction depends on large-scale online information dissemination.

3 Problem Formulation

With the collected data of students' campus card consumption, the friendship network $\mathcal{G}_f = (\mathcal{X}, \mathcal{V}_f)$ is constructed, where $\mathcal{X}$ is the set of nodes and $\mathcal{V}_f$ is the set of edges. For each node i in the network, $S_i = [s_{i1}, s_{i2}, ... s_{in}]$ retains node attributes. The d-dimensional embedded representation matrix $E \in R^{(|\mathcal{X}| \times d)}$ of $\mathcal{G}_f$ is obtained by using the mapping function $f(x)$ (where d means embedding spatial dimension). The resulting matrix $E \in R^{(|\mathcal{X}| \times d)}$ is used as the input of the three-layer neural network. We train the model to predict whether there will be an information dissemination relationship (ℓ_{st}) between two nodes s and t. Ultimately, we get the information dissemination network $\mathcal{G}_n(\mathcal{G}_n = (\mathcal{X}, \mathcal{V}_n))$, $\mathcal{V}_n$ is the set of ℓ_{st}.

Information Dissemination Network Generation Problem: Given the friendship network $\mathcal{G}_f$ and facial features data, psychological features data S, to generate the corresponding campus information dissemination network $\mathcal{G}_n$.

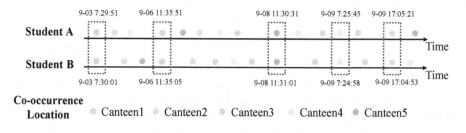

Fig. 2. Illustration of two students' co-occurrence.

4   Design of ANSWER

4.1   Construction of Friendship Network

Digital information management systems in 21st-century universities have dramatically improved the convenience of campus life for students. Temporal and spatial-based behavioral data provide us with the opportunity to discover potential social ties [3,23]. This paper utilizes students' campus card consumption records to construct a friendship network. Figure 2 explains what is co-occurrence. Student A and B co-occurs five times.

In our work, we focus on analyzing the consumption records of students in six different canteens. The process can be reformulated with more details as follows. Set the time interval to one minute [28], collect the consumption records of two students in the same canteen, and use the probability model to calculate the co-occurrence probability. The higher the probability is, the closer the relationship between friends is. The probability formula is calculated as [3]:

$$p(F|C_k) = \frac{P(F)P(C_k|F)}{P(C_k)} \approx \frac{1}{M}e^{klog\beta(N-1)+1} \tag{1}$$

where N represents geographic units, M individuals, joint visit probability β, prior probability $P(F) = \frac{1}{M-1}$, likelihood function $P(F) = p_1^k$, $p_1 = \beta + \frac{1-\beta}{N}$.

4.2   Processing of Node Attributes

Previous studies have shown that face images can be used to identify relationships [1]. To eliminate the effect of noise and obtain the vector representation of students' facial cognitive features, the convolutional autoencoder has been widely applied [18,26]. Convolutional autoencoder learns the sample features without labels and incorporates the properties of the convolutional neural network to complete unsupervised feature extraction. For the input layer x, it's corresponding i_{th} convolution is denoted as:

$$h^i = \sigma(x * \mathbf{W}^i + b^i), i = 1, 2, 3...k \tag{2}$$

$*$ is a convolution operation determined by context, and $\sigma(.)$ is the sigmoid function. And then the feature reconstruction process is represented as:

$$y = \sigma(\sum_{i \in H} h^i * \tilde{\mathbf{W}}^i + C) \tag{3}$$

$\tilde{\mathbf{W}}^i$ indicates filter weight, C indicates a bias vector. We use BP (Error Back Propagation) algorithm to optimize the loss function:

$$E(\theta) = \frac{1}{2n} \sum_{j=1}^{n} (x_j - y_j)^2 \tag{4}$$

Another attribute of nodes is psychological evaluation data, which is collected by questionnaires. There are some invalid, redundant, and missing data. First of all, for the invalid questionnaire, we will find the same participant and ask him to fill it out again. Then, the abnormal score data are detected by the box graph, and some missing data are filled by the median, and the unreasonable score data of students' individual tests are filled by mean. These data are recorded and encrypted corresponding to the student's ID, to protect their privacy.

4.3   Attributed Network Representation Learning (ANRL)

Through ANRL [30] of the deep neural network, node attributes, and network topology structure can be uninterruptedly integrated into the low-dimensional representation space. Therefore, in this paper, the network structure $\mathcal{G}_f$ and node attributes S of each node in the friendship network are input into the model to obtain the vector representation of each node.

The ANRL is mainly composed of the autoencoder and the Skip-gram model, the final optimization cost function is:

$$\mathcal{L} = \mathcal{L}_{sg} + \alpha \mathcal{L}_{ae} + \beta \mathcal{L}_{reg} \tag{5}$$

where $\mathcal{L}_{ae}$ is the loss function of the autoencoder and $\mathcal{L}_{sg}$ is the loss function of the Skip-gram. α is the hyper-parameter to balance the two losses and β is the l_2 parametric regularization factor.

The autoencoder is a neighbor-enhanced model that addresses the noise problems. The optimized loss function is:

$$\mathcal{L}_{ae} = \sum_{i=1}^{n} ||\hat{\mathbf{x}}_i - T(v_i)||_2^2 \tag{6}$$

where $\hat{\mathbf{x}}_i$ is the reconfigured output decoder. The reconstructed target $T(v_i)$ is the peculiarities of the target neighbor, which can be computed in two ways: weighted average of neighbor characteristics, $T(v_i) = \frac{1}{|\mathcal{N}(i)|}\sum_{j\in\mathcal{N}(i)} w_{ij}\mathbf{x}_j$ and median neighbor of element, $T(v_i) = \tilde{\mathbf{x}}_i = [\tilde{x}_1, \tilde{x}_2, ..., \tilde{x}_m], \tilde{x}_k = \text{Median}(w_{i1}\mathbf{x}_{1k}, w_{i2}\mathbf{x}_{2k}, ..., w_{i|\mathcal{N}(i)|}\mathbf{x}_{|\mathcal{N}(i)|k})$. $\mathcal{N}(i)$ is the neighbors of node v_i in the friendship network, and $\mathbf{x}_j$ is the feature vector of node v_i, including student facial cognition and psychological data, where w_{ij} indicates whether to weight or not.

The cost function $\mathcal{L}_{sg}$ for the Skip-gram model minimization is:

$$\mathcal{L}_{sg} = -\sum_{i=1}^{n}\sum_{c\in C}\sum_{-b\leq j\leq b, j\neq 0} \log p(v_{i+j}|\mathbf{x}_i) \tag{7}$$

where n indicates the total number of nodes, C is the collection of node sequences and b means the window size. $p(v_{i+j}|\mathbf{x}_i)$ is the conditional probability. Since the model extracts information about node attributes and global structure, it is computationally expensive and then we use negative sampling optimization [17]. For each specific node pair (v_i, v_{i+j}), the optimization goal is:

$$\log \sigma(\mathbf{v}_{i+j}^{'T} f(\mathbf{x}_i)) + \sum_{s=1}^{|neg|} \mathbb{E}_{v_n \sim P_n(v)}[\log \sigma(-\mathbf{v}_n^{'T} f(\mathbf{x}_i))] \tag{8}$$

$\sigma(.)$ is the sigmoid function, the value is $1/(1 + \exp(x))$. $\mathcal{L}_{reg}$ is formulated as:

$$\mathcal{L}_{reg} = \frac{1}{2}\sum_{k=1}^{K}(||\mathbf{W}^{(k)}||_F^2 + ||\hat{\mathbf{W}}^{(k)}||_F^2) \tag{9}$$

K is the number of layers of the encoder and the decoder. For the encoder, $\mathbf{W}^{(k)}$ is the k-th layer weighted matrix, and $\hat{\mathbf{W}}^{(k)}$ is the k-th layer weighted matrix of the decoder. The final representation of y_i^K extracts the node v_i attribute characteristics and network structure information.

4.4   Link Prediction

In this paper, the problem of reconstructing an information dissemination network is changed to a link prediction problem. We use ANRL to train the friendship network with node attributes, and obtain a low-dimensional representation of each node $\mathbf{E} \in R^{(|\mathcal{X}|\times d)}$, which is input in the deep learning model for training. In addition, we adopt the real information dissemination relationships as the labels for training. The three-layer neural network is selected as the information dissemination prediction model for the prediction task.

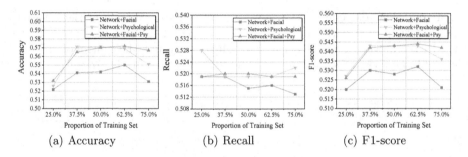

Fig. 3. Performance comparison of different inputs of ANSWER. The horizontal coordinate is the proportion of the training set.

5   Experiments

5.1   Dataset

We collect data from freshmen in the same major at a university in China. The data includes the facial image of students, psychological evaluations, and data in the information dissemination scene and students' campus card records. The students were notified about the use of these data and signed a consent form to confirm their participation in the research. They were noticed that these data were only used for academic research and were stored encrypted in encoded form, no intuitive information exists, and only relevant researchers could access.

The dataset includes three parts: real information dissemination data, student psychological level data, and campus card record. (A) In order to obtain the real information dissemination relationships of students, the questionnaire was adopted. Students need to answer: who will they voluntarily share the information with? Students need to write down 6–8 people. (B) The psychological level data of the students were collected by completing the Dominance Test Scale [6] and the Big Five Personality Scale [5]. There are different scoring standards for the scale, so the psychological characteristic data needs to be normalization processing before used. (C) Formally obtain student campus card information from the school's digital campus management department, and legally obtain facial photos of participating volunteers.

5.2   Prediction

Performance Comparison of Different Inputs of ANSWER: The student features include facial and psychological attributes. To verify the effect of facial and psychological attributes in the prediction task, we compare the network structure by including only facial or psychological attributes and both. Following that, we adopted three different combinations to learn different vector

representations of nodes, and finally completed the prediction of the information dissemination network. For the performance of ANSWER framework, we
use F1-score, Recall, and Accuracy to evaluate. Through the control of different
model inputs, the effectiveness of the ANSWER framework is proved.

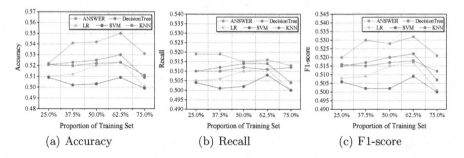

(a) Accuracy   (b) Recall   (c) F1-score

Fig. 4. Comparison of the performance of different prediction algorithms on friendship
networks combined with face attributes.

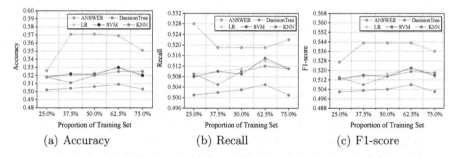

(a) Accuracy   (b) Recall   (c) F1-score

Fig. 5. Comparison of the performance of different prediction algorithms on friendship
networks combined with psychological attributes.

From Fig. 3, the prediction performance of the friendship network and facial
attributes on the information dissemination relationship is significantly inferior
to that of the friendship network combined with psychological attributes or all
three attributes. When the training ratio of the training set and the test set is
1:1, the predictive performance results of the friendship network combined with
psychological attributes are similar to that combined with the facial and psychological attributes. But then as the training set ratio increases, the combination
of all three is more effective. Through the input variant control of the ANSWER,
the network topology combined with the facial and psychological attributes is
more effective for information dissemination relationship prediction.

**Comparison of Differences in Results with Different Prediction
Algorithms:** The three-layer neural network in the ANSWER model is replaced

by Logistic Regression (LR), K-Nearest Neighbor (KNN), Support Vector Machine (SVM), and Decision Tree (DT) algorithms for information dissemination relationship prediction, and three different input combinations of the ANSWER framework are tested for comparison. The results are shown in Fig. 4, 5, 6.

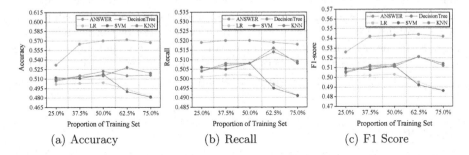

(a) Accuracy	(b) Recall	(c) F1 Score

Fig. 6. Comparison of the performance of different prediction algorithms on friendship networks combined with facial as well as psychological attributes.

★ The ANSWER has strong nonlinear fitting capabilities and multi-layer feature fusion with neural networks. The findings of all experiments indicate the outstanding performance of the ANSWER in predicting information dissemination relationships.

★ The KNN algorithm without feature fusion uses Euclidean Distance to calculate sample similarity, which depends largely on the number of selected neighbors, and the performance lower than ANSWER.

★ From Fig. 6, We found that LR and SVM results are much lower than when the network structure contains only one attribute. ANRL combines facial and psychological features into embedding vectors through nonlinear learning. But LR and SVM are linear models. When the eigenvector increases, they may not make full use of the embedding vector after being combined in a nonlinear manner.

★ As the training set proportion increases, the model is better trained. However, when it reaches 62.5%, we observe that the effect starts to gradually decrease. The reason may be the scarcity of sample data. As a result, the test set at this time is smaller, so the strengths and weaknesses of the model cannot be reflected more effectively.

6   Conclusion

In this paper, our goal is to solve the problem of campus information dissemination network generation. Based on the friendship network, combining the students' facial features and psychological features to create an attribute network, we use the friend attribute network to build an effective framework ANSWER for campus information dissemination relationship prediction. As far as we know, our work is the first to make predictions about campus information dissemination relationship. We compensate for this limitation in a novel way. And the results show that, compared with other predictive relationship algorithms, ANSWER composed of network representation learning and deep learning model fuses various features through nonlinear learning, which has outstanding performance in generating and forecasting campus information dissemination relationship. The sample volume for this paper is relatively small because of the privacy issues involved in the data and calling for more volunteers is very challenging. Future plans are to obtain larger datasets and to enhance the generalizability of the framework for other campus social relationships, such as academic cooperation networks.

References

1. Addington, J., Saeedi, H., Addington, D.: Facial affect recognition: a mediator between cognitive and social functioning in psychosis? Schizophr. Res. **85**(1–3), 142–150 (2006)
2. Campan, A., Cuzzocrea, A., Truta, T.M.: Fighting fake news spread in online social networks: actual trends and future research directions. In: IEEE International Conference on Big Data, pp. 4453–4457 (2017)
3. Crandall, D.J., Backstrom, L., Cosley, D., Suri, S., Huttenlocher, D.P., Kleinberg, J.: Inferring social ties from geographic coincidences. Proc. Nat. Acad. Sci. **107**(52), 22436–22441 (2010)
4. Duan, J., Luo, Y., Wang, Z., Huang, Z.: Semi-supervised cross-modal hashing with graph convolutional networks. In: Borovica-Gajic, R., Qi, J., Wang, W. (eds.) ADC 2020. LNCS, vol. 12008, pp. 93–104. Springer, Cham (2020). https://doi.org/10.1007/978-3-030-39469-1_8
5. Goldberg, L.R.: An alternative "description of personality": the big-five factor structure. J. Pers. Soc. Psychol. **59**(6), 1216 (1990)
6. Hamby, S.: The dominance scale: preliminary psychometric properties. Violence Vict. **11**(3), 199–212 (1996)
7. Horton, D., Richard Wohl, R.: Mass communication and para-social interaction: observations on intimacy at a distance. Psychiatry **19**(3), 215–229 (1956)
8. Hou, Y., Chen, H., Li, C., Cheng, J., Yang, M.C.: A representation learning framework for property graphs. In: Proceedings of the 25th ACM SIGKDD International Conference on Knowledge Discovery and Data Mining, pp. 65–73 (2019)
9. Kaur, R., Sharma, M., Taruna, S.: Privacy preserving data mining model for the social networking. In: 2019 6th International Conference on Computing for Sustainable Global Development, pp. 763–768 (2019)
10. Kong, X., Shi, Y., Wang, W., Ma, K., Wan, L., Xia, F.: The evolution of turing award collaboration network: bibliometric-level and network-level metrics. IEEE Trans. Comput. Soc. Syst. **6**(6), 1318–1328 (2019)
11. Kong, X., Shi, Y., Yu, S., Liu, J., Xia, F.: Academic social networks: modeling, analysis, mining and applications. J. Netw. Comput. Appl. **132**, 86–103 (2019)
12. Kramer, A.D., Guillory, J.E., Hancock, J.T.: Experimental evidence of massive-scale emotional contagion through social networks. Proc. Nat. Acad. Sci. **111**(24), 8788–8790 (2014)

13. Kumar, P.R., Wan, A.T., Suhaili, W.S.H.: Exploring data security and privacy issues in internet of things based on five-layer architecture. Int. J. Commun. Netw. Inf. Secur. **12**(1), 108–121 (2020)

14. Lerman, K., Ghosh, R.: Information contagion: an empirical study of the spread of news on digg and twitter social networks. In: Proceedings of the Fourth International AAAI Conference on Weblogs and Social Media, pp. 90–97 (2010)

15. Liu, J., et al.: Artificial intelligence in the 21st century. IEEE Access **6**, 34403–34421 (2018)

16. Liu, J., et al.: Shifu2: a network representation learning based model for advisor-advisee relationship mining. IEEE Trans. Knowl. Data Eng. 1 (2019)

17. Mikolov, T., Chen, K., Corrado, G., Dean, J.: Efficient estimation of word representations in vector space. In: Proceedings of Workshop at ICLR (2013)

18. Mirjalili, V., Raschka, S., Namboodiri, A., Ross, A.: Semi-adversarial networks: convolutional autoencoders for imparting privacy to face images. In: 2018 International Conference on Biometrics, pp. 82–89 (2018)

19. Mussumeci, E., Coelho, F.C.: Reconstructing news spread networks and studying its dynamics. Social Netw. Anal. Min. **8**(1), 1–8 (2018). https://doi.org/10.1007/s13278-017-0483-9

20. Pan, H., Guo, T., Bedru, H.D., Qing, Q., Zhang, D., Xia, F.: DEFINE: friendship detection based on node enhancement. In: Borovica-Gajic, R., Qi, J., Wang, W. (eds.) ADC 2020. LNCS, vol. 12008, pp. 81–92. Springer, Cham (2020). https://doi.org/10.1007/978-3-030-39469-1_7

21. Patton, G.C., et al.: Adolescence and the next generation. Nature **554**(7693), 458–466 (2018)

22. Perozzi, B., Al-Rfou, R., Skiena, S.: Deepwalk: online learning of social representations. In: Proceedings of the 20th ACM SIGKDD International Conference on Knowledge Discovery and Data Mining, pp. 701–710 (2014)

23. Sohail, A., Hidayat, A., Cheema, M.A., Taniar, D.: Location-aware group preference queries in social-networks. In: Wang, J., Cong, G., Chen, J., Qi, J. (eds.) ADC 2018. LNCS, vol. 10837, pp. 53–67. Springer, Cham (2018). https://doi.org/10.1007/978-3-319-92013-9_5

24. Tang, J., Qu, M., Wang, M., Zhang, M., Yan, J., Mei, Q.: Line: large-scale information network embedding. In: Proceedings of the 24th International Conference on World Wide Web, pp. 1067–1077 (2015)

25. Tasnim, S., Hossain, M.M., Mazumder, H.: Impact of rumors and misinformation on COVID-19 in social media. J. Prev. Med. Public Health **53**(3), 171–174 (2020)

26. Tewari, A., et al.: MoFA: model-based deep convolutional face autoencoder for unsupervised monocular reconstruction. In: Proceedings of the IEEE International Conference on Computer Vision Workshops, pp. 1274–1283 (2017)

27. Vosoughi, S., Roy, D., Aral, S.: The spread of true and false news online. Science **359**(6380), 1146–1151 (2018)

28. Yao, H., Nie, M., Su, H., Xia, H., Lian, D.: Predicting academic performance via semi-supervised learning with constructed campus social network. In: Candan, S., Chen, L., Pedersen, T.B., Chang, L., Hua, W. (eds.) DASFAA 2017. LNCS, vol. 10178, pp. 597–609. Springer, Cham (2017). https://doi.org/10.1007/978-3-319-55699-4_37

29. Zhang, D., et al.: Judging a book by its cover: the effect of facial perception on centrality in social networks. In: The World Wide Web Conference, pp. 2290–2300 (2019)

30. Zhang, Z., et al.: ANRL: attributed network representation learning via deep neural networks. In: Proceedings of the Twenty-Seventh International Joint Conference on Artificial Intelligence, pp. 3155–3161 (2018)
31. Zhao, Z., Zhang, S.: Research on influence of weak ties to information spreading in online social networks. In: 2013 13th International Symposium on Communications and Information Technologies, pp. 210–215 (2013)
32. Zhou, C., Liu, Y., Liu, X., Liu, Z., Gao, J.: Scalable graph embedding for asymmetric proximity. In: Proceedings of the Thirty-First AAAI Conference on Artificial Intelligence, pp. 2942–2948 (2017)

Twitter Data Modelling and Provenance Support for Key-Value Pair Databases

Asma Rani[1]($\boxtimes$), Navneet Goyal[2], and Shashi K. Gadia[3]

[1] Department of Computer Science, DBRAIT, PortBlair, India
asma.sags@gmail.com
[2] Department of Computer Science, BITS Pilani, Pilani, Rajasthan, India
goel@pilani.bits-pilani.ac.in
[3] Department of Computer Science, IOWA State University, Ames, USA
gadia@iastate.edu

Abstract. In Big Data environments, reliability of data plays an important role to determine trustworthiness of the outcomes of an analysis. Big data provenance ensures the reliability of data by providing details about the origin and historical paths of data. In recent years, the preponderance of big data and its applications are increasingly using Apache Cassandra due to its high availability and linear scalability. In this paper, we present a data provenance framework for Key-Value Pair Databases using the concept of Zero-Information Loss Database (ZILD). A large volume of real-time social media data is fetched from the Twitter's network through live streaming with the help of Twitter Streaming APIs, and then modelled in Apache Cassandra based on a Query-Driven approach. This framework provides efficient provenance capturing support for select, aggregate, update, and historical queries. We evaluate the performance of proposed framework in terms of provenance capturing and querying capabilities using appropriate query sets.

Keywords: Data provenance · Zero-Information Loss Database · Social media · Key-value pair · Provenance querying

1 Introduction

In present times, large volume of heterogeneous and unstructured data is generated every second via social media, scientific applications, workflows etc. This huge volume of data is stored and processed by different company's viz. Google, Facebook, and Amazon to provide services to the users or improving business, user experience etc. To provide such kind of services, applications are being developed which entrust this data, and can store and process huge volume of heterogeneous and unstructured data efficiently and reliably. For this, databases with high availability, high reliability and with ability to scale to large volumes in a distributed manner are becoming prevalent. Such databases are known as No-SQL (Not Only SQL) databases. There are over 150 different database products that belong to NoSQL family and Apache Cassandra is one of the most

© Springer Nature Switzerland AG 2021
M. Qiao et al. (Eds.): ADC 2021, LNCS 12610, pp. 87–98, 2021.
https://doi.org/10.1007/978-3-030-69377-0_8

popular ones. Apache Cassandra is used in application development [16,18] by Facebook, Twitter, Cloudkick, and Mahalo etc.

Data provenance [3,19] deals with the problem of identifying the source and derivation process of information. It plays an important role in applications where trustworthiness, privacy, and security of big data are major concerns [6]. In such big data applications, provenance information can be used to provide explanation or proof for query results. Capturing provenance for such applications is very challenging because of the high volume and unstructured data. A number of challenges are presented for provenance support in big data application by different authors [1,6,19] like automatic provenance capture, different granularity levels at which provenance needs to be captured, provenance capturing overhead, and analyzing data via querying provenance etc. A number of provenance models are existing in the literature which capture the whole system provenance [9,20,24], and provenance for workflows [8,13,17,22,23]. To the best of our knowledge, there is only one data provenance model for Cassandra, a key-value pair system [10]. Existing provenance model for Cassandra is suitable for a small database and is application specific. It is suitable for provenance of update queries only. There is no support for provenance of select and aggregate queries. Moreover, it uses the Thrift API, an older version of Cassandra Query, which makes expressing queries quite tedious.

In this paper, we propose a provenance framework based on the concept of Zero-Information Loss Databases [4,5,15] for Key-Value Pair Databases. One of the main contributions of the paper is the design and development of Query-Driven Zero-Information Loss Key-Value Pair Database (ZILKVPDB) which is used to efficiently store large volumes of live streaming Twitter data for efficient querying, maintaining updates, and to provide provenance capturing and querying support.

2   Related Work

Different approaches for data modeling in Apache Cassandra to perform efficient querying are proposed in [2,7,12,21]. A provenance data model for data intensive Map Reduce workflows was proposed in [8] to capture provenance using Kepler-Hadoop framework. A provenance framework, RAMP was proposed in [17,22] for Generalized Map and Reduce Workflows (GMRWs) using a wrapper based approach for provenance capturing and tracing. After this, HadoopProv [23] was introduced for provenance tracking in Map Reduce workflows, by treating provenance tracking in Map and Reduce phases separately. A big provenance framework, Milieu [24], was proposed for provenance collection and storage in unstructured or semi-structured format, for scientific applications/experiments as a workflow. Various change data capture (CDC) schemes were investigated in [14] for Apache Cassandra to track modifications in source data.

To satisfy the need of Big Data provenance, a provenance model for key-value pair system, was proposed in [10,11] to capture provenance information using provenance policies and querying on provenance information via resource expressions and a set of predefined operators. The proposed model was implemented on a small sized patient information system and used an older format of Cassandra Query using "thrift" approach that makes it difficult to write a query. From the available literature, it is evident that most of the existing provenance models are suitable to capture provenance for workflows at coarse-grained level only rather than fine-grained level. Secondly, some of them are not suitable for capturing provenance information either for a large data sets or for almost all types of query set. In this paper, we try to bridge this gap by designing an efficient data provenance framework on top of key-value pair database for capturing, storing and querying provenance information for different query set including select, aggregate, update, and historical queries on a large real-life twitter data sets.

3   Proposed Provenance Framework

In this paper, we propose a provenance framework which is built on the top of a Zero-Information Loss Key-Value Pair Database (ZILKVPDB). The concept of zero-information loss database was proposed in [15], and later used in provenance models [4,5]. The proposed provenance framework is suitable for applications which produce live streaming data. For illustration, we have modelled Twitter data in Apache Cassandra (Key-Value Pair Database). ZILKVPDB is capable to efficiently capture provenance information for various query sets including select, aggregate, update, and historical queries. The proposed framework also supports the querying of provenance information for historical data, tracing origin, and derivation history of result tuple of queries. Major steps involved in designing the framework are:

1. To perform live streaming of real-life Twitter dataset related to a specific event using Twitter Streaming API's.
2. To design an efficient Query-Driven Data Model for modelling Twitter Dataset in Key-Value Pair Database for efficient queries on twitter data.
3. To propose a provenance framework for Capturing and Storing Provenance information for Select Queries, Aggregate Queries, Historical Queries, and Update queries in Key-Value Pair Database based on ZILKVPDB.
4. To provide Provenance Querying support for querying historical data, tracing origin and derivation history of result tuples of queries.

3.1   Data Model Design

Twitter is one of the most popular social networking micro-blogging sites. In our proposed framework, Twitter Streaming APIs are used to retrieve continuously stream tweets and extract related information for designing a Query-Driven Key-Value Pair Data Model in Apache Cassandra. The proposed Data Model, as shown in Fig. 1, is based on frequent queries required to execute on Twitter dataset. The data model contains a keyspace named "NewTwitter_Keyspace" that consists of 20 Column Families. The various column names of these column families with their row keys are mentioned in Fig. 1.

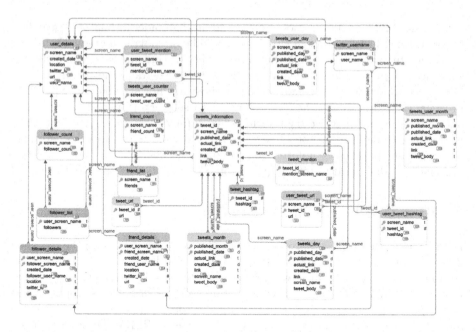

Fig. 1. Twitter data model in Cassandra

3.2 Zero-Information Loss Key-Value Pair Database

The proposed data provenance framework is based on Zero-Information Loss Key-Value Pair Database (ZILKVPDB) approach, to maintain all the updates without any information loss as provenance information. Provenance information about all the updates is captured in "update_provenance" column family. ZILKVPDB has been designed and implemented as per Algorithm 2 (Update ProvCassandra), explained in Sect. 3.3. We also propose Extended Cassandra Query Language (CQL) Constructs such as "all", "instance", "validon now" and "validon date" to support historical data queries as per Algorithm 3. Detailed explanation of these constructs with example queries are presented in Sect. 3.4.

3.3 Provenance Generation Algorithms

In this section, we present two provenance generation algorithms, one for select and aggregate queries, and other for update queries respectively with suitable example queries. Algorithm 1 (SelectProvCassandra) shows high level details of provenance generation for Select/Aggregate queries which takes a Select/Aggregate Query Q_S as input and returns the list of Provenance Paths (P) for all the result tuples of Q_S in the form of provenance path expressions i.e. "keyspace/columnfamily/rowkey(value)/columnname", and the updated Column Families viz. "select_provenance" , and "query_table". Demonstrations of Algorithm 1 with examples for select and aggregate queries are given below:

Algorithm 1 *SelectProvCassandra*: Provenance Generation for Select/ Aggregate Query

---

Input: Q_S (Select/Aggregate Query)

Output: Provenance_Paths Expressions (P) of all attributes in Result Tuples (T), Updated select_provenance and query_table Column Families

1: **Parse** Q_S
2: $KS, CF, PK, CN \leftarrow$ **Retrieve**   //Where KS=KeySpace, CF=Column_Family,
   PK=Primary Key of Column_Family, CN=Column Names in Q_S
3: $Q_{WS} \leftarrow$ **Rewrite** Q_S by adding PK
4: $T \leftarrow$ **Execute** Q_{WS}
5: **for all** $t \in T$ **do**
6:    $p \leftarrow$ Null
7:    **Obtain** v_{PK} , v_{CN} //Where V_{PK} =Values of Primary Key Columns,
      V_{CN} = Values of Column Names in Q_S, $v_{PK} \in V_{PK}$, $v_{CN} \in V_{CN}$
8:    $RK \leftarrow v_{PK}$ //Where RK is RowKey
9:    **for** i=1 to n **do**
10:      $p_i \leftarrow KS/CF/RK/C_i$
         //Where $C_i \in CN$ and C_i is non-key column, p_i is provenance path of C_i
11:      $p \leftarrow$ **Append** p_i // where p is list of provenance paths of C_i's in t
12:   **end for**
13:   $P \leftarrow$ **Append** p
14:   select_provenance $\leftarrow$ **Insert** Q_S, P, Current Date/Time
15: **end for**
16: query_table $\leftarrow$ **Insert** QueryId, Q_S, Current Date/Time
17: **End**

Example Query 1: *Display the location of user with Screen_Name Gagan4041.*
Cassandra Query 1: *select location from user_details where screen_name= 'Gagan4041';*

Result of above select query consists of two columns viz. LOCATION and LOCATION_PROVENANCE with value "India" and "[NewTwitter_ Keyspace /user_details/Gagan4041/location]" respectively. Here, the value under column LOCATION_PROVENANCE justify the query result i.e. "India". It explains that the value in result set is derived from keyspace: NewTwitter_Keyspace, column family: user_details, row key: Gagan40041, column: location.

Example Query 2: *Display the total no of tweets posted by a user "sunilthalia" on "08/10/ 2019".*
Cassandra Query 2: *select count(tweet_body) from tweets_user_day where screen_ name ='sunilthalia' and published_day=8 and published_date>= '2019-10-08' and published_date< '2019-10-09' group by screen_name allow filtering;*

The above query is an aggregate query to retrieve the total number of tweets posted by the user on a given day i.e. 7, along with comma separated list of provenance path expressions of all 7 tweets i.e. [NewTwitter_Keyspace/ tweets_user_day /sunilthalia-8-Tue Oct 08 11:37:56 IST 2019/tweet_body,New Twitter_Keyspace/tweets_user_day/sunilthalia-8-Tue Oct 08 11:40:12 IST 2019/ tweet_body, New Twitter_Keyspace/tweets_user_day/sunilthalia-8-Tue Oct 08 11:48:33 IST 2019/ ...] which justify the query result.

We propose Algorithm 2 (Update ProvCassandra) to design Zero-Information Loss Key-Value Pair Database (ZILKVPDB) which takes Update Query (Q_U) as input, performs update and returns provenance path expression of updated tuples along with updated Column Families viz. "update_provenance", and "query_ table" as provenance information. The provenance information about updates helps in historical data queries as well as provenance for standing queries. Illustration of Algorithm 2 is shown with Example Query 3 below:

Algorithm 2 *UpdateProvCassandra*: Provenance Generation for Update Query

---

Input: Q_U (Update Query)

Output: Provenance_Paths Expressions (P) of Updated Tuples (T),
        Updated update_provenance and query_table Column Families.

1: **Parse** Q_U
2: $KS, CF, PK, UCN, UCV, UCT$ ← **Retrieve**   //Where KS =KeySpace, $CF=Column\_Family$, $PK=Primary$ Key of Column_Family, $UCN=Update$ Column Name, $UCV = Update$ Column Value, $UCT = Updated$ Column Type
3: **if** Q_U contains **where Clause then**
4:    **Retrieve** V_{PK} ← **Parse where Clause**   //Where V_{PK} = Values of Primary Key Columns
5:    $RK ← V_{PK}$ //Where RK is RowKey
6:    Q_S ← **Generate** Select Query Corresponding to Q_U to obtain old value and
            WriteTime of UCN
7:    OCV , OCV_{WT} ← **Execute** Q_S //Where OCV = Old Column Value, OCV_{WT} =Old Column Value WriteTime
8:    $P ← KS/CF/RK/UCN$ //Where P is Provenance_Path Expression
9:    update_provenance ← **Insert** Q_U, P, OCV , OCV_{WT} , UCV, UCT , Current Date/Time
10: **else**
11:    Q_S ← **Generate** Select Query Corresponding to Q_U to obtain old value, write time   of UCN and value of Primary Key Columns
12:    T ← **Execute** Q_S
13:    **for all** $t \in T$ **do**
14:       **Set** $P =$ Null   //Where P is Provenance_Path Expression
15:       **Obtain** V_{PK} , OCV , OCV_{WT} ← t
16:       $RK ← V_{PK}$
17:       $P ← KS/CF/RK/UCN$
18:       update_provenance ← **Insert** Q_U, P, OCV , OCV_{WT} , UCV, UCT , Current Date/Time
19:    **end for**
20: **end if**
21: **Execute** ← Q_U
22: query_table ← **Insert** QueryId, Q_U, Current Date/Time
23: **End**

Example Query 3: *Update location of user with name "DDNewsAndhra".*
Cassandra Query 3: *update user_details set location = 'Andhra' where screen_name= 'DDNews Andhra';*

Above query performs update with new value in user_details column family and also capture old value before update along with its time of existence in "update_provenance" column family which makes our ZILKVPDB.

All the captured provenance information is stored in three different column families in Cassandra viz. "query_table", "select_provenance", and "update_provenance" for further analysis like auditing, error tracing, debugging, querying historical data etc.

3.4   Querying Provenance

The proposed provenance framework supports for querying provenance information for audit purpose, tracking all updates, and any other suitable application. The stored provenance information can be queried in two perspectives viz. 1. Justify the Query Result i.e. How any result tuple of select query is derived - querying provenance to know about source of information, 2. To track all the updates of any value - query provenance for historical data. To query the provenance information from perspective 1, framework supports to query "select_provenance" column family. From perspective 2, framework supports querying "update_provenance" column family to query provenance information about historical data. To know about all the queries executed till now and their time of execution, it can be retrieve with ease via "query_table" column family. In this section, we will illustrate the provenance querying support in our framework with some example provenance queries below:

Example provenance Query 1: *Explain how result tuple t1 of query q7 is derived.*

The above query will execute on "select_provenance" column family to retrieve the provenance path expressions of result tuple t1 of query q7 along with its time of execution. Here, provenance path expression of resultant tuple is [NewTwitter_Keyspace/ user_details/Gagan4041/location] and Time of query execution is "2019-12-16 05:02:34.266000+0000". This justifies that the source of required tuple is NewTwitter_Keyspace Keyspace, user_details Column Family, Gagan40041 Row Key and column name location at time 2019-12-16 05:02:34. 266000+0000. But, in case, if the source has been modified after query execution, still the original source can be devised via querying historical data. Our framework provides support for Extended Cassandra Query Language (CQL) constructs by introducing operators viz. "all", "instance", "validon now" and "validon date" to get historical data as well as current data. Algorithm 3 shows high level details of querying historical data using Extended Cassandra Query Language (CQL) constructs proposed in the framework.

Illustration of Algorithm 3 for querying provenance information for historical data with some example queries is presented below:

Example Provenance Query 2: *Display all the location updates of a specific user till now.*

Algorithm 3 *QueryProv_HistData*: Querying Provenance for Historical Data

---

Input: Query(Q), Operator(Instance or All), Time(Now or Date)
Output: Result Set(RS) of Q with given Operator and Time in Input

1: **Parse** Q
2: $KS, CF, PK, CN \leftarrow$ **Retrieve**   //*Where KS =KeySpace, CF=Column_Family,*
    PK=Primary Key of Column_Family, CN=Column Name
3: **if** $Operator = InstanceANDTime = Now$ **then**
4:    **Retrieve** $RS \leftarrow$ **Execute** Q on CF //*Where RS is ResultSet*
5: **else if** $Operator = InstanceANDTime = Date$ **then**
6:    $Q_P \leftarrow$ **Generate** Select Query Corresponding to Q and Time to obtain value of
      CN from UpdateProvenance(UP) Column Family
7:    **Retrieve** $RS \leftarrow$ **Execute** Q_P on UP
8:    **if** $RSisNull$ **then**
9:        **Retrieve** $RS \leftarrow$ **Execute** Q on CF
10:    **end if**
11: **else**
12:    $Q_P \leftarrow$ **Generate** Select Query Corresponding to Q and Time to obtain value
      of CN from UpdateProvenance(UP) Column Family
13:    **Retrieve** $RS \leftarrow$ **Execute** Q_P on UP
14:    **Retrieve** $RS \leftarrow$ **Execute** Q on CF
15: **end if**
16: **End**

Extended CQL Query: *select all location from user_details where screen_name = 'MemeBaaaz' validon now;*

The above query will be parsed first to retrieve operator and time which is "all" and "now" respectively in this case. Thus, simple CQL query will be executed to retrieve all location updates of the given user in "user_details" and "update_provenance" column families. The result of above query is shown in Table 1 along with time of existence of each location update.

Example Provenance Query 3: *Display all the location updates of a specific user till 23/10/2019 9:50AM.*

Extended CQL Query: *select all location from user_details where screen_name= 'MemeBaaaz' validon 2019-10-23 09:50:16;*

The above query will generate only first two rows of Table 1 in its result set because the location "Mumbai" is valid after time '2019-10-23 09:50:16 'given in query.

Table 1. Example provenance query 2 result

LOCATION	VALID_FROM
Meme Ki Duniya, India	Wed Oct 02 13:33:27 IST 2019
Kolkata	Wed Oct 23 08:20:18 IST 2019
Mumbai	2019-12-17 10:22:22.0

Example Provenance Query 4: *Display the current location of a specific user.*
Extended CQL Query: *select instance location from user_details where screen_name='MemeBaaaz' validon now;*

The above query will generate current location of user as "Mumbai" that is valid from "2019-12-17 10:22:22.0".

Example Provenance Query 5: *Display the location of a specific user on date 23/10/2019 8:22:16AM.*
Extended CQL Query: *select instance location from user_details where screen_name='MemeBaaaz' validon 2019-10-23 08:22:16;*

The above query will generate location of user at 23/10/2019 8:22:16AM as "Kolkata" which is valid from "Wed Oct 23 08:20:18 IST 2019" to "2019-12-17 10:22:21.0".

4   Results and Discussions

To evaluate the performance of proposed provenance framework, all the experiments are performed on a single node Apache Cassandra Cluster on Intel i7-8700 processor @ 3.20 GHz with 16 GB RAM, and 1TB Hard Disk. Apache Cassandra version 3.11.3 has been used for the experiments. All the queries are executed on Cassandra database using Cassandra Query Language (CQL). Real-life Twitter data set consists of around 1.8 lakh twitter user's, 1.7 lakh users friends, 1.2 lakh user's followers, and their other information like tweets posted, personal information etc., is used for the experiments.

To perform experimental analysis of provenance capture, we have executed a set of 25 different queries including both data retrieval and data update queries. Average execution times of all the queries are mentioned in milliseconds (ms) as shown in Fig. 2, where the performance overhead for the queries with provenance capturing mechanism to capture and store provenance information is very minimal as compared to queries without provenance capturing mechanism except a few queries like Q10, Q11, and Q20. As the proposed framework captures and store provenance information of each result tuple in the result set of a query, which increases the execution time with increase in number of result tuples. Therefore, the query Q10 with provenance capturing support is taking longer execution time, as it is producing larger number of result tuples in its result set.

The proposed framework also provides provenance support for aggregate queries. In Fig. 2, the queries Q6, Q11, Q12, and Q20 are aggregate queries those are using aggregate functions on some input values based on the predicate given in that query. Although, the framework efficiently captures and stores provenance information for aggregate queries such as query Q6, and Q12, yet it takes more execution time for the queries where aggregation is performed on large number of tuples in input data set such as query Q11, and Q20.

The provenance capturing for update queries are also supported by the proposed framework. We have executed a set of data update queries to capture and store their provenance information in "update_provenance" column fam-

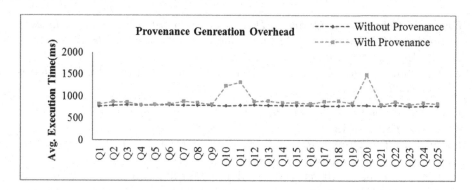

Fig. 2. Query performance without and with provenance

ily. These provenance informations can be used for historical data queries, and queries executed in past at any specific time as presented in previous section. Like select, and aggregate queries, the proposed framework support provenance capturing and storing for update queries efficiently with a minimal execution time overhead, see query Q22, Q23, Q24, and Q25 in Fig. 2.

Table 2. Provenance performance overhead (ms)

	Update queries	Select queries	Aggregate queries
Without Provenance	782	788	794
With Provenance	838	866	1124

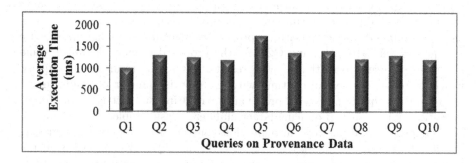

Fig. 3. Provenance querying

The average query execution times in millisecond with and without provenance support for "Update", "Select", and "Aggregate" Queries are presented in Table 2. We found that proposed framework is efficient for "update", "select", and "aggregate" queries in terms of execution overhead viz. 1.07, 1.10, and 1.42 percents respectively. We have executed a set of provenance queries as explained Sect. 3.4 the performance analysis of querying provenance information is shown in Fig. 3. It shows that the proposed framework efficiently supports provenance querying for both source/origin tracing and historical data queries.

5 Conclusion and Future Work

In this paper, we have developed a Data Provenance Framework to capture provenance information for real-life live streaming twitter data set modelled in Zero-Information Loss Key-Value Pair Database (ZILKVPDB). The proposed framework has the potential to capture provenance information for query set which includes select queries, aggregate queries, and update queries. It also supports to capture provenance information for historical queries using data versioning support in ZILKVPDB. The proposed framework is efficient in terms of average execution time for capturing and storing the provenance information for select queries, and update queries. However, a small execution overhead is measured for some aggregate queries where aggregation is performed on a large number of input tuples. Framework also provides support for querying provenance information for historical data queries, as well as to trace the source of result tuples of select/aggregate queries. In future work, framework will be further extended for distributed data on different nodes in a cluster.

References

1. Chacko A., Kumar S.D.M.: Big data provenance research directions. In: IEEE Region 10 Conference (TENCON), pp. 651–656 (2017)
2. Chebotko A., Kashlev A., Lu, S.: A big data modeling methodology for Apache Cassandra. In: IEEE International Congress on Big Data, pp. 238–245 (2015)
3. Imran, A., Agrawal, R.: Data Provenance. In: Schintler, L.A., McNeely, C.L. (eds.) Springer Proceeding of Encyclopedia of Big Data. Springer, Heidelberg (2017). https://doi.org/10.1007/978-3-319-32001-4_58-1
4. Rani, A., Goyal, N., Gadia S.K.: Data provenance for historical queries in relational database. In: ACM Compute, pp. 117–122 (2015)
5. Rani, A., Goyal, N., Gadia, S.K.: Efficient multi-depth querying on provenance of relational queries using graph database. In: ACM Compute, pp. 11–20 (2016)
6. Glavic, B.: Big data provenance: challenges and implications for benchmarking. In: Rabl, T., Poess, M., Baru, C., Jacobsen, H.-A. (eds.) WBDB -2012. LNCS, vol. 8163, pp. 72–80. Springer, Heidelberg (2014). https://doi.org/10.1007/978-3-642-53974-9_7
7. Bermbach, D., Mullery, S., Eberhardt, J., Tai, S.: Informed schema design for column store-based database services. In: 8th International Conference on Service-Oriented Computing and Applications (SOCA), pp. 163–172 (2015)
8. Crawl, D., Wang, J., Altintas, I.: Provenance for mapreduce-based data-intensive workflows. In: 6th Workshop on Workflows in Support of Large-scale Science (WORKS11), pp. 21–30 (2011)
9. Ghoshal, D., Plale, B.: Provenance from log files: a BigData problem. EDBT/ICDT 13, 290–297 (2013)
10. Kulkarni, D.: A provenance model for key-value systems. In: 5th USENIX Workshop on the Theory and Practice of Provenance (TaPP), Article No. 12 (2013)

11. Kulkarni, D.: A fine-grained access control model for key-value systems. In: 3rd ACM Conference on Data and Application Security and Privacy (CODASPY 2013), pp. 161–164 (2013)
12. Ramesh, D., Kumar, A.: Query driven implementation of twitter base using Cassandra. In: IEEE International Conference on Current Trends toward Converging Technologies, pp. 1–4 (2018)
13. Hondo, F., et al.: Data provenance management for bioinformatics workflows using NoSQL database systems in a cloud computing environment. In: IEEE International Conference on Bioinformatics and Biomedicine, pp. 1910–1915 (2017)
14. Schmidt, F.M., Geyer, C., Schaeffer-Filho, A., Debloch, S., Hu, Y.: Change data capture in NoSQL databases: a functional and performance comparison. In: 20th IEEE Symposium on Computers and Communication (ISCC), pp. 562–567 (2015)
15. Bhargava, G., Gadia, S.K.: Relational database systems with zero information loss. IEEE Trans. Knowl. Data Eng. **5**(1), 76–87 (1993)
16. Wang, G., Tang, J.: The NoSQL principles and basic application of Cassandra model. In: IEEE International Conference on Computer Science and Service System (CSSS 2012), pp. 1332–1335 (2012)
17. Park, H., Ikeda, R., Widom, J.: RAMP: a system for capturing and tracing provenance in mapreduce workflows. VLDB Endow. **4**(12), 1351–1354 (2011)
18. Mahmood, K.: Performance comparison of NOSQL database Cassandra and SQL server for large databases. Indep. Stud. Res. Comput. **14**(2), 21–25 (2016)
19. Senellart, P.: Provenance in databases: principles and applications. In: Krötzsch, M., Stepanova, D. (eds.) Reasoning Web. Explainable Artificial Intelligence. LNCS, vol. 11810, pp. 104–109. Springer, Cham (2019). https://doi.org/10.1007/978-3-030-31423-1_3
20. Agrawal, R., Imran, A., Seay, C., Walker, J.: A layer based architecture for provenance in big data. In: IEEE International Conference on Big Data, pp. 1–7 (2014)
21. Hernandez, R., Becerra, Y., Torresa, J., Ayguade, E.: Automatic query driven data modelling in Cassandra. Elsevier Procedia Comput. Sci. **51**, 2822–2826 (2015)
22. Ikeda, R., Park, H., Widom, J.: Provenance for generalized map and reduce workflows. In: 5th Biennial Conference on Innovative Data Systems Research (CIDR 11) (2011)
23. Akoush, S., Sohan, R., Hopper, A.: HadoopProv: towards provenance as a first class citizen in MapReduce. In: 5th USENIX Workshop on the Theory and Practice of Provenance (TaPP 13), Article No. 11 (2013)
24. Cheah, Y., Canon, R., Plale, B., Ramakrishnan, L.: Milieu: lightweight and configurable big data provenance for science. In: IEEE International Congress on Big Data, pp. 46–53 (2013)

Analyzing Tweets to Understand Factors Affecting Opinion on Climate Change

S. Mohith⬤, Jackson I. Jose⬤, Sonia Khetarpaul⬤, and Dolly Sharma$^{(\boxtimes)}$⬤

Department of Computer Science and Engineering,
Shiv Nadar University, NCR, Greater Noida, India
{ms207,jj779,sonia.khetarpaul,dolly.sharma}@snu.edu.in

Abstract. Climate change is a topic that is frequently debated on social media. A vast majority in the debate cite scientific evidence to recognize the existence of a man-made climate change and its impacts on environment as well as society. The opinion of the masses is critical to dealing with various issues arising due to climate change, such as global warming. In this work, we study people's opinion on climate change and analyze the data to identify the common topics which garner discussion. Our aim is to analyze the dataset, explore the popular belief of a region and then derive the possible explanation in terms of different factors. This analysis could help us in determining the extent to which different factors affect people's opinion. By building sentiment analysis models, performing topic modelling and using other appropriate technologies, we can visualise the sentiment pattern to understand the factors affecting them.

Keywords: Climate change · Data mining · Twitter Sentiment Analysis · Topic modelling

1 Introduction

Climate change was scientifically recognized in the early nineteenth century, but studies related to it remained in its infancy for a long time. It was only in the 1990s that many scientists agreed upon greenhouse gases and global warming. Climate change experts believe that if we continue the current carbon emission rate, then by the next century the Earth would be unfit for human habitation. World Bank reports (Hsieh et al. 2018) brings into light the threats on agriculture, water levels and coastal regions when the world would be 4 °C warmer. It also brought to discussion the unprecedented heat waves, severe drought, and major floods in many regions, with serious impact on ecosystems (Bank 2006). The scientific community has come to a unanimous conclusion that immediate changes are necessary and that inaction would cost us dearly. Political parties often exploit people's polarity on climate change, where lobbyists often spread misinformation to gain profits simply by instilling doubts around the science that proves climate change is real. The main avenue for them to do this is through the mainstream media, holding panel discussions and debates on the topic (Smith

© Springer Nature Switzerland AG 2021
M. Qiao et al. (Eds.): ADC 2021, LNCS 12610, pp. 99–110, 2021.
https://doi.org/10.1007/978-3-030-69377-0_9

2017). Another research studied difference between the groups responding on either side of the debate and analyzed their responses (Hodges and Stocking 2016). (Kim and Cooke 2018) analyzed tweets on climate change and ocean acidification before and after USA's withdrawal from Paris Climate Agreement. They observed that Trump's decision brought increased attention to climate change and people started expressing political opinions related to climate change on Twitter and concluded that social media could be an effective medium to share information and opinions. The effect of ocean acidification due to increasing amounts of CO2 produced from man made sources, will result into change of carbonate ion concentration in the ocean surface water in major oceans around the world (Feely et al. 2009). (Le et al. 2017) studied the 2016 USA presidential announcements on social network platforms and identified that campaign announcements spiked people's attention. However, they claimed that twitter cross talk was focused more on criticizing opposite party.

Due to the growing popularity of social networks, these portals have become a very rich source of data. Twitter is one of the largest social networking portals in the world, generating huge amounts of data every day. Twitter's question of "What's happening?" captures the primary objective of the portal well, i.e. to share what is happening around oneself. Online social network platforms also enabled researchers to conduct surveys on diverse population in large numbers in a very short amount of time. Given the imminent threat, we identified that it is necessary to check empirically whether there is any sort of correlation between people's sentiment on climate change and their political, social or economic background. This study would help to identify the flow of misinformation to the population, and their susceptibility to it.

The mindset and perspective of the people plays a really important factor in making such a collective effort successful. Due to the growing popularity of the internet, blogs, social media, these have become a very rich source of data. The data generated by these portals can help analyze the different factors that shape people's opinion. By doing so, the government, companies and other entities can make better decisions to make their efforts more successful. The key contributions of this paper are:

1. The use of Latent Dirichlet Allocation to identify the most debated and discussed topics in climate change across various regions. This allows us to derive insights from our data by correlating it with empirical evidence.
2. A comparison between the sentiments across two different year on how they varied across regions, the key events that occurred during those years which could have influenced the sentiments.
3. Visualisation of data at every point providing intuitive insights of sentiments across regions and time and also allowing us to picture a high level view of the variation in sentiments.
4. Correlation of the empirical data such as political, geographical factors with the trends observed in our data.

2   Literature Review

Kirilenko et al. (Kirilenko and Stepchenkova 2014) explored public discourse on climate change on Twitter on a larger scale. They discussed the relationship between geography, major news events and how central topics of discussion change over time across different countries. They also examined which organizational sources of information on climate change people in different countries generally referenced. They observed that discourse on climate change is significant in industrially developed countries of Europe, Australia, North America, India, and the Philippines. With a higher percentage of tweets coming from the urban areas of the US and England. They also observed significant temporal variability, with a substantially higher number of messages being tweeted during morning hours, and the first four days of the week. Their research describes what is said on climate change in a given place at a given time. They also commented upon how generally prominent people lead the debate on climate change.

Other researchers, who analyzed twitter data, classified the tweets into two categories, subjective and objective groups. They then further classified subjective tweets into those representing the group of tweets expressing concerning climate change and called for action, and others in the negative group that do not believe in climate change and called it a hoax. They detected a connection between abrupt changes in negative sentiments at times of extreme weather situations (An et al. 2014). 5.7 million tweets from USA, UK, Canada and Australia were analyzed and classified into 5 different classes - Real, Hoax, Cause, Impact and Action (Jang and Hart 2015). Pearce et al. collected tweets related to IPCC and classified them into four different categories: Supportive, Unsupportive, Neutral and Non- Tweeters (Pearce et al. 2014). They also analyzed tweets from hashtags related to science, politics, geographical discussions, societal changes and new technology to understand the trends in people who tweet on the topic and people who get tagged. (Segerberg and Bennett 2011) analyzed the ecology of a protest on social networks. The following three points were studied in the process, namely, process of intersection in the protest ecology in the network, the process of embedding or filtering of messages and reflection of changes in ecology.

Chen et al. collected 2000 tweets on climate change and labelled them as deniers and non-deniers. Next step was tokenization, normalization followed by vectorization. Further they used Deep learning to analyze temporal fluctuations in public opinion on climate change (Chen et al. 2019).

Dahal et al. used sentiment analysis and topic modeling using tweets containing spatial information and compared discussion on climate change across countries. They used LDA for topic modeling and a combination of VAD and SR for sentiment analysis (Dahal et al. 2019). This study uses a different methodology as compared to other studies but has certain limitations, such as, messages not written in English, missing context, filtered data (Dahal et al. 2019).

Sapul et al. compared topic modeling such as k-means clustering, LDA with a proposed clustering algorithm CLOPE where they found keywords and hashtags together find more meaningful topics (Sapul et al. 2017). Singleton et al.

performed sentiment analysis on tweets from coastal regions of USA and identified that southeast coast has smallest participation in climate change discussions (Singleton et al. 2019).

Studying 6000 climate change tweets, Veltri et al. performed (i) thematic analysis to create four clusters related to awareness/action, causes/consequences, policies and energy, (ii) semantic network analysis to identify various issues being discussed and (iii) text classification. However, the limitations of this study were that only English tweets within a period of seven days were analyzed (Veltri and Atanasova 2017).

Walter et al. analyzed the interaction patterns of scientists on social networks with people such as other scientists, citizens, journalists, politicians, etc. It was observed that they have intense discussions with other scientists, they use neutral language with citizens and express strong negative emotions while talking to media and politicians (Walter et al. 2019).

3   Experimental Methods

3.1   Data Collection

Over 350k tweets were crawled using the API provided by Twitter. The use of hashtags by Twitter users provides a way to query only those tweets which have a high probability of being relevant to the study. We crawled the commonly used climate change related hashtags such as '#climatechange', '#climate', '#climateaction', '#artic', '#globalwarming', '#unfcc', '#parisaccord', '#parisagreement', '#actionclimate'. A dataset by Kaggle, "Twitter Sentiment Analysis on Climate Change" was also used. This dataset had about 45k tweets, each Tweet labelled with their respective polarities. The polarities where divided into four classes, 2, 1, 0, −1, signifying news/facts, call for action/believing in climate change, neutral opinion and those who question the legitimacy of the subject respectively.

3.2   Data Preprocessing

The dataset obtained from Kaggle had truncated tweets, hence we crawled those tweets using their IDs. In our use case, cleaning the data was very important because of the usage of trending slangs, emoticons, etc. The steps we followed to clean the data are as follows:

1. Convert Kaggle dataset from HTML encoding to UTF-8 encoding.
2. Remove all user mentions and hashtags (begins with @ and) and links, as they do not contain any relevant information.
3. Tweets that are Retweets have "RT" appended at their start, which might skew our results, so we remove it.
4. Remove all stop words and lemmatize all tokens in every tweet.
5. Remove all tweets whose number of words is below 3 after all the steps above.

After performing these steps, about 350k tweets were left for the next step i.e, classification.

3.3   Classification Models

Here the tweet gets classified into two classes depending on whether it contains climate change opinion or not. Many models were employed as explained below.

Logistic Regression predicts the probability of an outcome that can only have two values. The curve is constructed using the natural logarithm of the "odds" of the target variable, rather than the probability.

Support Vector Machine is a discriminative algorithm which works by finding the best hyper-plane with maximum margin which separates the dataset provided. *Naive Bayes Classifier* is a probabilistic classifier which assumes that all the features present independent to each other, much similar to Bayes's theorem on which it is based. A key advantage of Naive Bayes Classifier is that it runs in linear time, hence performing experiments become much quicker. *Ensemble Classifier* is an ensemble learning method which uses multiple decision trees as its basis. Decision trees tend to over-fit the training data and have low bias and high variance, but when used in a Random Forest, multiple trees trained over different parts of the training data, the overall variance is reduced. This increases the performance significantly.

We have used a bag of words model for the vector representation of tweets. Here, the presence of a word impacts the result, but not the order. We calculated Term Frequency and Inverse Document Frequency (TF-IDF) for each term in our data set. We trained several different classifiers such as, Logistic Regression, Naive Bayes Classifier, Support Vector Machine Classifier, Random Forest Classifier and Balanced Bagging Classifier (Joachims 1998) (Dahal et al. 2019; Joachims 1998; Liaw and Wiener 2002; Ye and Ye 2020). To improve the accuracy, we classified the data in a two-step process. We first built a classifier to separate out all tweets that classify as class 2, i.e. news or facts, as they do not convey opinion of people.

$$TF = \frac{f}{w} \tag{1}$$

where: f = frequency of word in document w = number of words in document

$$IDF = \log \frac{N}{df_i} \tag{2}$$

N = Total number of documents in the corpus, df_i = Number of documents with the words.

The most notable performance is that of Random Forest Classifier (accuracy of 88%). We also created an Ensemble Model (Dietterich 2000). which uses multiple machine learning algorithms in a specific combination to improve the predictive performance. The performance analysis of various classifiers are summarized in the following tables (Table 1).

Next, tweets which were classified into −1 and 1, representing denying the existence of climate change and believing climate change is real respectively. As shown in Table 2, Support Vector Machines showed the best performance with

Table 1. Performance of classification models - segregate news and opinions

Classifier	Class	Precision	Recall	F1-score	Support
Logistic regression	0	0.94	0.80	0.86	7515
	1	0.49	0.79	0.60	1855
Naive bayes	0	0.83	0.99	0.91	7515
	1	0.87	0.19	0.31	1855
Ensemble classifier	0	0.90	0.96	0.93	7515
	1	0.79	0.56	0.65	1855
Random forest classifier	0	0.88	0.98	0.93	1507
	1	0.86	0.54	0.66	1855

Table 2. Performance of classification models - sentiment classification

Classifier	Class	Precision	Recall	F1-score	Support
Logistic Rrgression	−1	0.54	0.66	0.59	1358
	1	0.89	0.83	0.86	4614
Naive bayes	−1	0.87	0.39	0.54	1358
	1	0.85	0.98	0.91	4614
Support vector machine	−1	0.78	0.52	0.63	1358
	1	0.87	0.96	0.91	4614

an accuracy of about 85.9% when compared with others. The other factors that led to selecting SVM over others are as follows (Joachims 1998):

1. They handle high dimensional data without over-fitting.
2. Tweets have a limitation of 240 characters per tweet, which makes each data vector very sparse. Such conditions are better handled by SVM.
3. Text categorization is generally linearly separable, and the basic idea behind an SVM is to find the hyper plane that separates the categories.

3.4   Topic Modeling

Climate change is a very wide subject, and it assembles discussion in its various different facets such as politics, policy, economy or even just the overall weather. To find the different topics within tweets, we performed topic modelling using Latent Dirichlet Allocation (LDA) Ye and Ye (2020), an unsupervised clustering algorithm used to discover topics that may occur in a collection of document. The algorithm sees documents as comprising of various topics and each topic having its own set of words and then associates each document with a certain topic. We have clustered the tweets into ten different topics.

4   Result and Discussion

The top five countries in terms of number of tweets in our dataset were United States, United Kingdom, Canada, Australia and India in the same order. Figure 3 shows the percentage of negative tweets. The disparity could be attributed to the better internet penetration in these countries. Since we only gathered tweets in English, the Non-English speaking countries like Japan, China and others are not very well crawled as it can be seen from the map (Fig. 1).

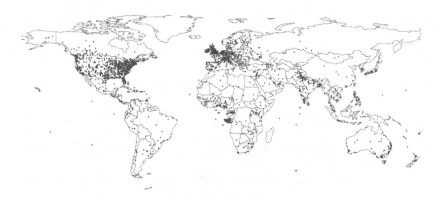

Fig. 1. Sentiments demanding actions

4.1   Opinion Shapers: Ideologies

Political leaders and their ideologies can have a big influence on people. In June 2017, U.S. President Donald Trump announced his intention to withdraw United States of America from the Paris Agreement (an agreement within the United Nations Framework Convention on Climate Change (UNFCCC), dealing with

Fig. 2. Denial sentiment

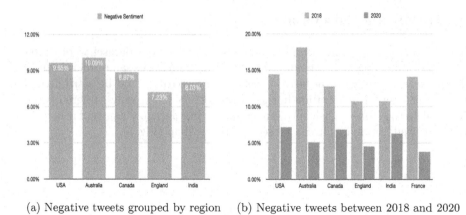

(a) Negative tweets grouped by region   (b) Negative tweets between 2018 and 2020

Fig. 3. Representation of negative tweets

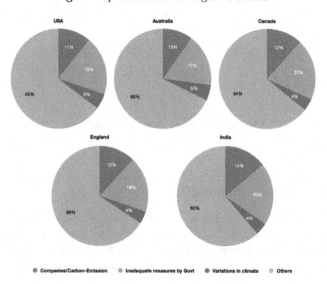

Fig. 4. Proportion of topics in the tweets: Result of LDA

greenhouse-gas-emissions mitigation, adaptation, and finance, signed in 2016) and ever since then, many policies that are contrary to the Paris Agreement are already put in place in USA. Mr. Trump has often pursued the interests of industries such as coal and car manufacturing at the expense of the environment in his pursuit of jobs growth. He has lifted restrictions on coal plants emissions and approved permits for the huge Keystone XL oil pipeline. He has rolled back the Obama administration's pollution standards for vehicles, even taking on California for wanting to impose their own separate – tougher – ones. Trump, clearly a populist has openly admitted that he does not believe in climate change. Trump supporters are bound to be influenced by his flow and his ideology, which may explain the rise in the number of climate change denialist found in US. Our data set aligns with the time Trump's popularity grew significantly, i.e. 2016 till present. According to Cody et al. (2015), in the United States climate change is a topic that is heavily politicised; the words "deny", "denial", and "deniers" are used more often in tweets containing the word "climate" (Fig. 2).

Australia is the second largest exporter of coal after Indonesia. It plays a big part in their economy. The government has approved the construction of a new coal mine, which would be the biggest in the world. Scott Morrison, the current Australian Prime Minister did not attend the UN Climate Action Summit held on September 2019, and the Australian government also stopped payments to the Green Climate fund. The Liberal Party of Australia, the current ruling party of Australia has members who continue to openly cast their doubt about climate change in their interviews. A spokesman for Mr. Scott Morrison said: "Australia has already outlined our policies to tackle climate change including cutting our emissions by 26–28% and investing directly into climate resilience projects through our regional partners" (Murphy 2019). However, the decision to construct the coalmine says otherwise. It is clear politically, Australia does benefit with lax policies (Goodman 2020).

4.2   Opinion Shapers: Leadership

The Prime Minister of Canada, Justin Trudeau is a vocal proponent for fighting climate change, who introduced carbon tax in 2018. However, he has faced criticism for breathing life into Canada's oil industry. Indeed, the day after the government declared a climate emergency; it approved a multi-billion-dollar oil pipeline expansion. Although, the revenue from here are being used for green projects. Due to this, the net greenhouse emissions in Canada has decreased ever so slightly under Justin's administration. Combine this with overall 40% land areas covered by forest, the climate conditions in Canada is quite better than other places. Such steps by the political leaders can convince the existence of global warming. This might explain the difference in the percentage of the sentiment of class −1 when compared to US, 8.87% vs 9.65%, despite them being similar in economic front, and sharing such a huge common border.

The former prime minister of UK Theresa May set a goal of net-zero emissions by 2050. Boris Johnson's (the current PM) father was a committed environmentalist who has campaigned for faster action on climate change. His influence is apparent in Boris Johnson's endorsement of various campaigns. As we can see from the pie chart in Fig. 4, UK has very few number of people who deny (and those who are skeptic/neutral) climate change among the top five countries. It maybe so because of the leader's belief and the policies (UNFCCC Paris Agreement, zero-GHG emission by 2050). The UK government has allotted more budget to fight climate change than what was allotted before indicating its commitment towards the 2050 goal.

4.3   Opinion Shapers: Media

"The greater the quantity of media coverage of climate change, the greater the level of public concern". Studies indicate that advertising interests and editors have always challenged journalists' abilities to adequately report on climate change issues. Instead of climate change stories, editors often prefer more sensational topics that garner higher ratings and approval with advertisers (Park 2018).

At a time when civilization is accelerating towards disaster, climate silence continues to reign across the bulk of the US news media. Especially on television, where most Americans still get their news, the brutal demands of ratings and money work against adequate coverage of the biggest story of our time. Many newspapers, too, are failing the climate test. The IPCC landmark report from 2017 that had some eye-opening facts/evidences supporting climate change was covered by just 22 of the 50 top media houses. In this aspect, the Canadian media is better. In late September, hundreds of thousands of Canadians took to the streets across the country to demand more from their governments on climate change. It was one of the largest mass protests in Canadian history. It was also a sign, many in the environment movement believed, of Canada's climate-change coming of age. Climate Change was chosen in a survey of reporters and editors across the country as the 2019 Canadian Press News Story of the Year.

Huge variations in climate change, its effects/evidence is not been covered by the mainstream media as viciously as they cover other subjects. We could take the example of India, which has about 8% class −1 tweet, which is quite a lot considering the severity of the matter. In India most media houses are pro establishment and are generally more concerned about debates in topics such as whether a person is anti- national or not? Is India faring better than Pakistan or not? etc. It is almost as if media houses do not want to cover more of climate change. One possible explanation for this is that the more the media houses cover these issues, the more the people start protesting with the government for its decisions like approving permits for oil pipeline or taking down pollution standards. If that happens, the companies involved could suffer huge losses and, in some cases, reduce job opportunities. So, it could be in the interest of the government to lobby for such industries, disregarding its impact on the environment, for immediate gain. It is also notable that despite having such a large user based, the proportion of tweets from India about climate change is far less.

5   Conclusion

In this study a variety of advanced tools were used to obtain the results and empirical evidence was used to explain the reason behind the observed results. Climate Change is one of the biggest challenges of our time and identifying the problem is only half the story. We can see that actions from political entities trigger large scale response from the people. It is crucial that these entities take correct decisions. Although we've tried to cover a majority of important events, these are not all. Every small incident at every level of the society plays an important role in shaping the opinions of the people. Twitter provides a reliable overview into people's life and the factors in their surroundings that shape their opinion. We also need to consider other popular platforms and languages to be able to understand the opinions better.

References

An, X., Ganguly, A.R., Fang, Y., Scyphers, S.B., Hunter, A.M., Dy, J.G: Tracking climate change opinions from Twitter data. In: KDD (2014)

Bank, W.: Turn down the heat: climate extremes, regional impacts, and the case for resilience. Modern Casting (2006)

Blei, D.M., Ng, A.Y., Jordan, M.I.: Latent Dirichlet allocation. J. Mach. Learn. Res. (2003). https://doi.org/10.1016/b978-0-12-411519-4.00006-9

Chen, X., Zou, L., Zhao, B.: Detecting climate change deniers on Twitter using a deep neural network. In: ACM International Conference on Proceeding Series Part F1481, pp. 204–210 (2019). https://doi.org/10.1145/3318299.3318382

Dahal, B., Kumar, S.A.P., Li, Z.: Topic modeling and sentiment analysis of global climate change tweets. Soc. Netw. Anal. Min. 9(1), 1–20 (2019). https://doi.org/10.1007/s13278-019-0568-8

Dietterich, T.G.: Ensemble methods in machine learning. In: Kittler, J., Roli, F. (eds.) MCS 2000. LNCS, vol. 1857, pp. 1–15. Springer, Heidelberg (2000). https://doi.org/10.1007/3-540-45014-9_1

Feely, R.A., Doney, S.C., Cooley, S.R.: Ocean acidification: present conditions and future changes in a high-CO2 world. Oceanography (2009). https://doi.org/10.5670/oceanog.2009.95

Goodman, J.: What is Australia doing to tackle climate change? [WWW Document]. BBC News (2020)

Hodges, H.E., Stocking, G.: A pipeline of tweets: environmental movements' use of Twitter in response to the Keystone XL pipeline. Env. Polit. 25, 223–247 (2016). https://doi.org/10.1080/09644016.2015.1105177

Hsieh, S.Y., et al.: Classifying protein specific residue structures based on graph mining. IEEE Access 6, 55828–55837 (2018). https://doi.org/10.1109/ACCESS.2018.2872496

Jang, S.M., Hart, P.S.: Polarized frames on "climate change" and "global warming" across countries and states: evidence from Twitter big data. Global Environ. Change (2015). https://doi.org/10.1016/j.gloenvcha.2015.02.010

Joachims, T.: Text categorization with support vector machines: learning with many relevant features. In: Nédellec, C., Rouveirol, C. (eds.) ECML 1998. LNCS, vol. 1398, pp. 137–142. Springer, Heidelberg (1998). https://doi.org/10.1007/BFb0026683

Kim, S.C., Cooke, S.L.: Environmental framing on Twitter: impact of Trump's Paris Agreement withdrawal on climate change and ocean acidification dialogue. Cogent Environ. Sci. 4, 1–19 (2018). https://doi.org/10.1080/23311843.2018.1532375

Kirilenko, A.P., Molodtsova, T., Stepchenkova, S.O.: People as sensors: mass media and local temperature influence climate change discussion on Twitter. Global Environ. Change (2015). https://doi.org/10.1016/j.gloenvcha.2014.11.003

Kirilenko, A.P., Stepchenkova, S.O.: Public microblogging on climate change: one year of Twitter worldwide. Global Environ. Change (2014). https://doi.org/10.1016/j.gloenvcha.2014.02.008

Feldman, S., Lavelle, M.: Donald Trump's Record on Climate Change [WWW Document]. Inside Climate News (2020)

Le, H., Boynton, G.R., Mejova, Y., Shafiq, Z., Srinivasan, P: Bumps and bruises: mining presidential campaign announcements on Twitter. In: HT 2017 - Proceedings of the 28th ACM Conference on Hypertext and Social Media, pp. 215–224 (2017). https://doi.org/10.1145/3078714.3078736

Liaw, A., Wiener, M:. Classification and Regression by RandomForest (2002)

Newman, T.P.: Tracking the release of IPCC AR5 on Twitter: users, comments, and sources following the release of the Working Group I Summary for Policymakers. Public Underst. Sci. **26**, 815–825 (2017). https://doi.org/10.1177/0963662516628477

NOAA: National Climate Report - May 2020 [WWW Document]. National Centers for Environmental Information (2020)

Park, D.J.: Environ. Assess. Manag. (2018). https://doi.org/10.1002/ieam.2011

Pearce, W., Holmberg, K., Hellsten, I., Nerlich, B.: Climate change on Twitter: topics, communities and conversations about the 2013 IPCC Working Group 1 report. PLoS One (2014). https://doi.org/10.1371/journal.pone.0094785

Sapul, M.S.C., Aung, T.H., Jiamthapthaksin, R.: Trending topic discovery of Twitter Tweets using clustering and topic modeling algorithms. In: Proceedings of the 2017 14th International Joint Conference on Computer Science and Software Engineering, JCSSE 2017 (2017). https://doi.org/10.1109/JCSSE.2017.8025911

Segerberg, A., Bennett, W.L.: Social media and the organization of collective action: using twitter to explore the ecologies of two climate change protests. Commun. Rev. (2011). https://doi.org/10.1080/10714421.2011.597250

Singleton, S., Kumar, S.A.P., Li, Z: Twitter analytics-based assessment: are the United States coastal regions prepared for climate change. In: Proceedings of the International Symposium on Technology and Society, 2018-November, pp. 150–155 (2019). https://doi.org/10.1109/ISTAS.2018.8638266

Smith, J.: Climate change: scientific evidence and the industry of denial. Missouri Rev. (2017). https://doi.org/10.1353/mis.2017.0049

Veltri, G.A., Atanasova, D.: Climate change on Twitter: content, media ecology and information sharing behaviour. Public Underst. Sci. **26**, 721–737 (2017). https://doi.org/10.1177/0963662515613702

Ye, N., Ye, N.: Naïve Bayes classifier. In: Data Mining (2020). https://doi.org/10.1201/b15288-3

Cody, E.M., Reagan, A.J., Mitchell, L., Dodds, P.S., Danforth, C.M.: Climate change sentiment on Twitter: an unsolicited public opinion poll. PLoS One **10**(8), e0136092 (2015). https://doi.org/10.1371/journal.pone.0136092

Optimal Placement of Taxis in a City Using Dominating Set Problem

Saurabh Mishra$^{(\boxtimes)}$ and Sonia Khetarpaul

Shiv Nadar University, Noida, India
{sm609,sonia.khetarpaul}@snu.edu.in

Abstract. Mobile application based ride-hailing systems, e.g., DiDi, Uber have become part of day to day life and natural choices of transport for urban commuters. However, the pick-up demand in any area is not always matching with the supply or drop-off request in the same area. Urban planners and researchers are working hard to balance this demand and supply situation for taxi requests. The existing approaches have mainly focused on clustering of the spatial regions to identify hotspots, which refer to the locations with a high demand for pick-up requests. In our study, we determined that if the hotspots focus on the clustering of high demand for pick-up requests, most of the hotspots pivot near the city center or two-three spatial regions, ignoring the other parts of the city. In this work, we proposed a method, which can help in finding a local hotspot to cover the whole city area. We proposed a dominating set problem based solution, which covers every part of the city. This will help the drivers looking for near-by next customer in the region wherever they drop their last customer. It will also reduce the waiting time for customers as well as for a driver looking for next pick-up request. This would maximize their profit as well as help in improving their services.

Keywords: Ride-hailing system · Hotspots detection · Location based service · Dominating set problem application

1 Introduction

Taxis are important parts of an urban transport system and mobile-based taxi hailing applications have made them the most preferred mode of transport for intra-city travel. Ride-hail wait time is the period between the moment when a rider hits "request" to when a ride-hail vehicle arrives. Ride-hail wait time plays an important role in a customer's experience of a trip. One of the main goals of a taxi hailing service is to reduce the ride-hail wait time for every customer; various studies show this improves customer experience.

In recent years, there has been an increase in human mobility leading to an exponential rise to the number of vehicles especially mobile based taxi hailing services that provide ease of commuting. This has also led to a rise in the number of institutions and researchers focusing on urban transport and its related

© Springer Nature Switzerland AG 2021
M. Qiao et al. (Eds.): ADC 2021, LNCS 12610, pp. 111–124, 2021.
https://doi.org/10.1007/978-3-030-69377-0_10

fields like profit maximization, route recommendation, traffic analysis, hotspot detection and pollution control measures.

Mobile application-based ride hailing service easily detect the pick-up and drop-off points in a spatial region. This brings a researcher's interest in detecting the hotspots. In the application of taxi demand analysis, hotspots [4] are the places of more than usual occurrence; that is, the places with high density of demand. Most of the previous hotspot detection and analysis related works [1,3] have discussed the clustering of the areas with high demand of pick-up requests. Most of the cities have some areas which are near to the city center or prominent places (like monuments, Market, business hub) and they have higher demand for taxi pickups while rest of the city has more or less a normal distribution of requests. Most of the recent researches are focused on clustering of those area while the other areas remain ignored. Practically, a taxi driver who has just dropped a passenger in a less prominent area cannot drive back every time to the city center to pick his or her next passenger. Instead he can wait for sometime or drive to a local hotspot which is nearer to his current drop-off location.

(a)   (b)

Fig. 1. Optimal Placement of taxis (a) Existing approach (b) Proposed model

Example: Figure 1 shows an example, where all the taxis waiting near the hot-spots detected on the basis of maximum pick-up request clustering. In our approach, we determine the local hot-spots across the city using road network graph and historical data of taxi pick-up request. This will help in finding the local hot-spots covering each part of the city. Focusing on these issues, our aim is to:

- Reduce the cruising time of the drivers, that also maximizes their revenues,
- Find the optimal number of taxis in a city that maximizes the revenue per taxi, and
- Maximize the coverage of taxi services in a city with the optimum resource allocation.

2   Literature Review

Identification of taxi hot-spots is relatively a recent and upcoming field of research [6–15]. Many researchers have been working on identifying the taxi hot-spots. The authors in [2] have proposed a simple Dijkstra-based algorithm for approachable kNN query on moving objects for the ride-hailing service, which considers the occupation of objects. They further improved its efficiency by applying a grid-based Destination-Oriented index for occupied and non-occupied moving objects.

In [3], the authors have applied density-based spatial clustering techniques to study the similarity in distribution pattern of pick-up and drop-off hotspots. In [4] the authors mined historical data to predict demand distributions with respect to different contexts of time, weather, and taxi location for predicting taxi demand hotspots. Yu et al. in [1] have proposed a conditional generative adversarial network a long short-term memory structure (LSTM-CGAN) model to capture the spatio-temporal distribution of taxi hotspots.

In many of the existing approaches, the researchers have applied various clustering algorithms, like K-means [9], DBSCAN algorithm [8, 15–17], fuzzy clustering [19] or taxi-data mining algorithms, such as the density based hierarchical clustering method [12], to identify taxi pickups locations. Those clustering-based models mainly focused on spatial features of historical data to understand the taxi requirements. In order to understand the taxi demands more accurately many researchers have explored the temporal properties [18, 20].

Due to recent growth in deep learning techniques, some Researchers have applied these techniques for traffic predictions, such as short-term traffic flow [22, 23], real-time traffic speed [24], and passenger-demand for real-time ride service [25, 26]. Dominating sets are one of the most discussed topics in graph analysis. It is widely used for finding the most influential nodes in a graph for a communication networks, social networks and road networks [27, 28]. He et al. used dominating set problem for quality improvement in wireless sensor network [29], she proposed a neural network model to find the minimum weakly connected dominating sets (WCDS) in a wireless sensor network.

As discussed, most of the researchers have used different clustering and deep learning techniques for the hot-spot detection and analysis. In this paper, we are using the dominating set problem-based solution for detecting the local hotspots that reduce the cruising time for drivers, maximize their revenues, help to find the optimal number of taxis in a city, and also maximize the coverage of taxi services in a city.

3   Context of the Problem

In a spatial region, a taxi driver's priority is to find the next customer at the earliest after dropping his current customer. While waiting for his next trip and halting at some place it is desirable for a driver to minimize his arrival time to the next customer. To reduce this arrival time the driver can be placed around a reference position known as 'hotspot'. That hotspot can be detected by analysing his last drop location, time and request history, etc. This analysis will be helpful for taxi drivers to decide where to move after dropping their present customer and where to halt so that it minimizes the time to pick-up the next customer.

3.1 Preliminaries

Definition 1 (Dominating Set). A subset D of a the vertex set $V(G)$ of a graph G is said to be the dominating set if every vertex not in D is adjacent to at least one vertex in D. If the subgraph $<D>$ induced by D is connected it is called *Connected Dominating set* and if this set is minimal then it is called as *Minimum Connected Dominating Set (MCDS)* [27].

Definition 2 (k-hop Dominating Set). In this paper, we used the concept of dominating set to find the most important/influential nodes in a road network graph. We modified the traditional concept of dominating set and defined the k-hop dominating set as:

A k-hop dominating set is the minimum number of nodes that cover the whole graph and each selected node covers all the nodes at k-hop distance. So, every node from the dominating set will be at least k-hop and at max $2k$-hop distance away from the other nodes in the dominating set. Other nodes have maximum k-hop distance from a dominating set node as every dominating set node covers all nodes up to k-hop distance. Section 5 of this paper discusses the effect on different metrics by varying the value of k, this helps in finding the optimal value of k.

Definition 3 (Road Network Graph). The road network graph $G(V,E)$ is a planar graph, with each road segment as an edge(V) of the graph, and the point at which two or more roads intersect, are represented as the vertex(E) of the graph G.

Definition 4 (Driver). Each taxi driver is represented by W and it has some attributes which are represented by:

ID_W: Unique identity number of the each driver W, l_W: Location of the driver W, ST_W: Time of the day at which driver W starts working, B_W: Current status(Busy/Not-busy) of the driver W, TD_W: Distance covered during trips by the driver W, ID_W: Distance covered by the driver W for next customer search, NT_W: Total number of trips covered by the driver W.

When a taxi request is generated, the system searches the nearest drivers using their location and their status (busy/not-busy). If a driver is free and his location matches with the taxi request location according to the system, his attributes get updated accordingly.

4 Methodology

In this section, we present the algorithms to address the problem of optimal placement of taxis in a city to improve the availability of taxis for customers and to maximize the revenue. The road network of a city is a planar graph. Using the available taxi data-set and road network data-sets, we can determine the demands for taxi arising at each node per unit time in a city. After determining the demands per node, we are generating a weighted graph and evaluating the maximum weighted minimum dominating set of nodes in the city. This model uses historical data to predict the next request generating nodes or the most prominent point for drivers to stay.

This work proposes the following four algorithms:

- To find all the neighbours of a node we proposed a function *NeighbourSearch* (Algorithm 1) which is used in Algorithm 2 and 3.
- Algorithm 2 and 3, for determining the dominating set.
- Algorithm 4 represents the workflow, and uses Algorithm 2 and 3 for task assignment to the drivers.

Algorithm 1: Algorithm for search function *NeighbourSearch*

Input: : (a) Road Network Graph *G(V,E)* (b) Target node *'root'* from the graph *G(V,E)* (c) *k* value
Output: List of nodes under *k-hop* distance of *'root'*
```
 1: list S = [ ]              //an empty array list for visited nodes in each iteration
 2: set Q = {root}            //a set of nodes that behaves like a queue
 3: while k != 0 do
 4:    for each element in Q do
 5:       vertex v = pop leftmost element from Q
 6:       for each neighbour i of vertex v do
 7:          if neighbour i not in S then
 8:             add i in S
 9:             add i in Q
10:          end if
11:       end for
12:    end for
13:    k = k-1
14:    return S
15: end while
```

4.1   Neighbour Search

To determine the dominating set of a planar graph, first, we introduced a function to find all the nodes which are under k-hop distance from the selected node. Algorithm 1 represent the pseudo code for function *NeighbourSearch*, it takes three arguments, the road network graph, the targeted node and the *k* value up to which the neighbouring nodes have to be added. An array list S and a set of nodes Q is initiated. A while loop will iterate for each *k* value and add the neighbours at each *k* value in the list S and replace the node in set Q with its neighbours. To find the all neighbours up to *k-hop* distance the *while loop* will iterate *k* times. This will return *list S* that contains target and its neighbouring nodes up to *k-hop* distance. The complexity of this algorithm is *O(V)* as this algorithm using a BFS search method.

4.2 k-hop Dominating Set Algorithm

Algorithm 2 depicts the *k-hop Dominating Set* formation on the basis of number of requests originating on each node. It initiates an empty list '*DS*' and a empty dictionary *freq*. A dictionary is a collection which is unordered, changeable and indexed tuples, each tuple is a set of key: value pairs. In this algorithm dictionary *freq* stores nodes as keys and the number of requests on respective node as value. In line 3–5, *freq* appends each node with respective 'total number of requests' originating on that node calculated using historical data. Line 6–15 uses this data from *freq* to find the node with maximum request originating and selects this node along with its neighbours up to *k-hop* distance using *NeighbourSearch*. In each iteration a node with maximum value got selected and that node and its neighbours got deleted from *freq*. When *freq* get empty, we get the dominating set *DS* for the graph. The complexity of this algorithm is $O(V_2)$, as for each element in the dominating set, function *NeighbourSearch* is called and each neighbour is checked to be deleted or not.

Algorithm 2: Algorithm for *k-hop Dominating Set*

Input: : (a) Road Network Graph $G(V,E)$ (b) Taxi request data history
Output: List of nodes from Dominating Set '*DS*'
1: list DS = [ ]   //*an array list for adding dominating set node at each iteration*
2: dictionary freq = { } //*an empty dictionary data structure*
3: **for** every vertex v of Graph G **do**
4:     calculate no. of request *freq(v)* on vertex v and add *(v,freq(v))* in a dict *freq*
5: **end for**
6: **while** freq is not empty **do**
7:     maxnode = select node with maximum value from freq
8:     add maxnode to DS
9:     nbr = NeighbourSearch(G, maxnode, k value)
10:    **for** j in nbr **do**
11:        **if** j in freq **then**
12:            delete freq(j)
13:        **end if**
14:    **end for**
15: **end while**
16: return DS

4.3 Modified k-hop Dominating Set Algorithm

Next, we present the Algorithm 3 which is updated version of Algorithm 2. *k*-hop Dominating Set algorithm greedily selects the node that has maximum number of request among the available nodes at each iteration. Their is no consideration of requests appearing on the neighboring nodes, so in Algorithm 3 (Modified *k*-hop Dominating Set algorithm) tries to remove this frailty to improve the dominating set results (as shown in Fig. 2). Modified *k*-hop Dominating Set algorithm considers consolidated number of requests on each node including its all neighbours from *k*-hop distance.

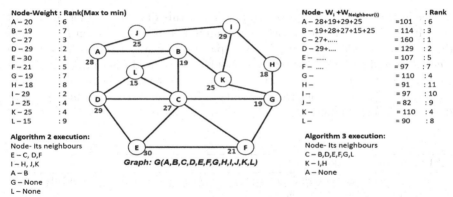

Node-Weight : Rank(Max to min)
A – 20   : 6
B – 19   : 7
C – 27   : 3
D – 29   : 2
E – 30   : 1
F – 21   : 5
G – 19   : 7
H – 18   : 8
I – 29   : 2
J – 25   : 4
K – 25   : 4
L – 15   : 9

Node- W_i +W_Neighbour(i)   : Rank
A – 28+19+29+25   =101   : 6
B – 19+28+27+15+25   = 114   : 3
C – 27+.....   = 160   : 1
D – 29+.....   = 129   : 2
E – .....   = 107   : 5
F – ....   = 97   : 7
G –   = 110   : 4
H –   = 91   : 11
I –   = 97   : 10
J –   = 82   : 9
K –   = 110   : 4
L –   = 90   : 8

Algorithm 2 execution:
Node- Its neighbours
E – C, D,F
I – H, J,K
A – B
G – None
L – None

Algorithm 3 execution:
Node- Its neighbours
C – B,D,E,F,G,L
K – I,H
A – None

Graph: G(A,B,C,D,E,F,G,H,I,J,K,L)

k-hop Dominating Set : E, I, A, G, B   *Modified k-hop Dominating Set : C, K, A*

Fig. 2. Execution of Algorithm 2 and 3 to find the *k-hop Dominating set* with *k = 1*

Algorithm 3: Algorithm for *Modified k-hop Dominating Set*

Input: : (a) Road Network Graph $G(V,E)$ (b) Taxi request data history
Output: List of nodes from Dominating Set *'DS'*
1: list DS = [ ]   //*an array list for adding dominating set node at each iteration*
2: dictionary freq = { }   //*an associative array to store unordered key-value pairs*
3: dictionary freq' = { }   //*an associative array to store unordered key-value pairs*
4: **for** every vertex v of Graph G **do**
5:     calculate no. of requests *freq(v)* on vertex v and add *(v,freq(v))* in a dict *freq*
6: **end for**
7: **for** every element i in the freq **do**
8:     freq'(i)=0
9:     nbr = NeighbourSearch(G, i, k value)
10:    **for** j in nbr **do**
11:       **if** j in freq **then**
12:          freq'(j) = freq'(j)+ freq(i)
13:       **end if**
14:    **end for**
15: **end for**
16: **while** freq' is not empty **do**
17:    maxnode = select node with maximum value from freq'
18:    add maxnode to DS
19:    nbr = NeighbourSearch(G, maxnode, k value)
20:    **for** j in nbr **do**
21:       **if** j in freq' **then**
22:          delete freq'(j)
23:       **end if**
24:    **end for**
25: **end while**
26: return DS

In this algorithm, first we append each node with respective 'total number of requests' originating on that node calculated using historical data in *freq*. Then, using the *NeighbourSearch* we calculated the total weight of each node along with its neighbours up to *k-hop* distance and saved this data in another dictionary *freq'*. In line 16–25, the node with maximum weight is selected from *freq'* and in the graph, this node and its neighbours up to *k-hop* distance selected. Each time a node selected, this node and its neighbours get removed from *freq'*. Finally, we get the *DS* when *freq'* gets empty which means that all the nodes of the graph *G* got covered. The complexity of this algorithm is $O(V_2)$, as for each vertex of the graph *G(V,E)*, function *NeighbourSearch* is called.

Algorithm 4: Task assignment to the drivers

Input: : (a) Road Network Graph *G(V,E)* (b) Taxi request data (c) List of Drivers
Output: Detailed information of all drivers
 1: *function CheckRequest(graph, Request, ListofDrivers)*
 2: Find near-by idle drivers using KNN-search
 3: **if** driver available at the requesting node **then**
 4:     assign driver this request
 5:     update the driver information in the driver list
 6: **else**
 7:     assign driver this request
 8:     travel to the requesting node
 9:     update the driver information in the driver list
10: **end if**
11: **return** updated driver information
12: **for** every request **do**
13:     driver information = CheckRequest(graph,Request,List of Drivers)
14:     **if** if driver location node is not is DS **then**
15:         move driver to DS
16:         update the driver information
17:     **end if**
18: **end for**

Figure 2 shows the execution of Algorithm 2 and 3. To find the *k-hop Dominating Set* using Algorithm 2 and 3 with $k = 1$, that means for each node, in a graph *G(A, B, C, D, E, F, G, H, I, J, K, L)*, if the node is picked then all its neighbours at 1 hop distance will also be picked. Finally it will return the set of nodes that cover the whole graph.

4.4   Task Assignments

Task assignment in a spatial region is one of the most important part in this system and assigning the right job(request) to the right person(driver) is a complex problem. Algorithm 4 depicts the work flow of the system. On arrival of every request, a *CheckRequest* function gets invoked. With this *CheckRequest* using the location of the request we check the nearest available taxi using *R-tree* based *KNN* search in that spatial region. If the driver is not busy and available at exact node (line 3–5) he accepts the request and his attributes get updated, else he travels to the desired location and his attributes get updated accordingly.

When the driver reaches at the destination of his current trip, he would stop in case that destination point is a node from a dominating set (proposed by Algorithm 2 or 3). Otherwise the driver would drive to the nearest node from the dominating set.

5   Experiment Setup

In this section, we experimentally evaluated the proposed algorithms to find the dominating set of the New York city road network graph [30]. Then, we analysed drivers' placement in that dominating set and studied its effect on using different parameters like varying the number of drivers and size of dominating set. In our experiments, we have used NY city road network graph [30] and NY taxi dataset [31]. The taxi data has 3.8 million taxi requests and 99% of these requests are extended from 40.5 N to 41.0 N latitude and -74.2 E to -73.5 E longitude. We cleaned the road map data and in the NY-city road map, we had 20,700 nodes and 33,000 edges. The NY-city taxi dataset has attributes: date-time, pickup location and drop-off location, and there are five to six thousand requests generated per day approximately. We considered taxi drivers data with different attributes: Id, location, status (busy/not-busy) and date-time. We used python libraries for our simulation and the experiments performed on Intel(R) Core(TM) i5-7500 CPU @3.40 GHz and 8 GB RAM.

5.1   Size of the Dominating Set Varying k-Value

The k-value is defined as the minimum distance between any two nodes of the dominating set. If $k = 5$, it means any two nodes in the dominating set have at least five nodes between them and maximum ten. Higher the k-value means more coverage of neighbours by each node. Figure 3 shows the effect of varying the number of nodes for NY-City road network graph. We can observe that the increasing value of k reduces the dominating set size exponentially.

5.2   Varying the Number of Drivers

Number of taxis available in a city plays an important role for taxi service providing companies. It can also be helpful in city traffic and other related infrastructure, which are however, beyond the scope of this work. In this section, we analysed the effect of number of taxis with and without the application of our proposed method, and observe how our approach makes a difference.

Figure 4 (a, c, e) and 4 (b, d, f) depict the performance varying the number of drivers using Algorithm 2 and 3, respectively. First, we analysed the number of drivers who got the work, that means at least one trip during the day. We can observe in Fig. 4(a) and 4(b) that our recommendations help more drivers to get the requests and with increasing numbers almost 80% of drivers get at least one ride. Figure 4(c) and 4(d) show an average of total number of taxi requests completed during the day. We can also observe from the figure that as the number of drivers increases, the number of request completed are also increased up to more then 80% of the total requests subsequently. Figure 4(e) and 4(f) show distance travelled by a driver to pick the next customer per request. We can see that with increasing number of drivers, distance travelled is decreased per request. This will save both resources and time and help in quality improvement.

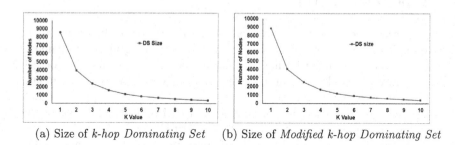

(a) Size of *k-hop Dominating Set*   (b) Size of *Modified k-hop Dominating Set*

Fig. 3. Size of the dominating set for different k-values

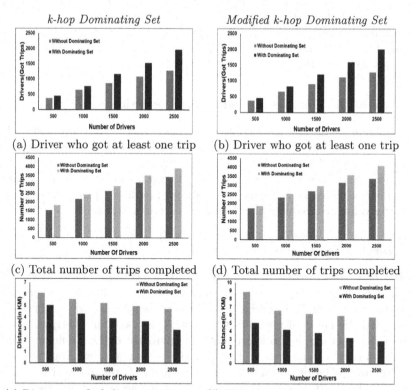

(a) Driver who got at least one trip   (b) Driver who got at least one trip

(c) Total number of trips completed   (d) Total number of trips completed

(e) Distance to find the next customer (f) Distance to find the next customer

Fig. 4. Effect of varying the number of drivers.

5.3 Varying the K-Value

Figure 5 (a, c, e) and 5 (b, d, f) depict the performance varying the k-value i.e., number of hops in a dominating set, using Algorithm 2 and 3, respectively. As the *k-value* increases the number of nodes in the dominating set decreases. First, we analysed the number of drivers who got the work, that means at least one trip during the day. We can see in Fig. 5(a) and 5(b), our recommendations helps more drivers to get requests as the area covered by each node of a dominating set is increased, which also increases the maximum reach of the drivers. With the increase in k-value, the size of dominating set gets decreased and each node covers more number of neighbouring nodes.

Figure 5(c) and 5(d) shows an average of total number of taxi request completed during the day with increasing k-value. We can observe from the figure as the k-value increases the number of requests completed are increased and reached up to more than 80% of the total requests. Figure 5(e) and 5(f) show distance travelled by a driver to pick the next customer per request, we can see with the increasing k-value, distance travelled also increases per request as the area (number of nodes) increases for each node in the dominating set. So this

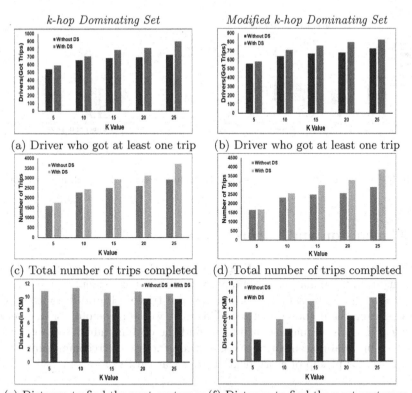

(a) Driver who got at least one trip   (b) Driver who got at least one trip

(c) Total number of trips completed   (d) Total number of trips completed

(e) Distance to find the next customer (f) Distance to find the next customer

Fig. 5. Effect of varying k-value (size of the dominating set).

factor becomes a trade-off between the distance travelled and the total number of taxi requests a driver gets to maximize his profit. To find an optimal k-value that can maximize the profit for a driver and his company is also a challenge.

From the experiments, it is observed that our approach provides more number of trips even with fewer number of taxis available in the region. Therefore, this approach is helpful in determining the optimal number of taxis, in a city, to get the maximum revenue in the same demand scenario.

5.4   Comparison with Other Clustering Algorithms

We compared our proposed algorithms (K-hop and Modified K-hop) with the well-known clustering algorithms viz. K-means, Agglomerative hierarchical clustering (AHC) and Density-Based Spatial Clustering of Applications with Noise (DBSCAN) algorithm, used in [4] for finding the hotspots in a city. The results are depicted in Fig. 6(a) and 6(b). The results show the effect of varying the number of drivers on the total number of trips, and distance covered by the driver to pick next customer. It is observed that our proposed algorithms outperform in both the cases.

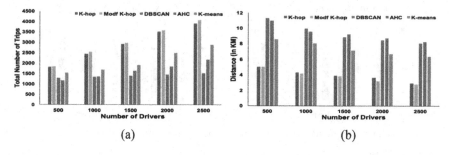

Fig. 6. (a) Total Number of Trips and (b) Distance covered to pick next customer

6   Conclusion and Future Work

In this paper, we have proposed algorithms for the optimal placement of taxis in a city using dominating set. From the results, it is evident that there is a significant improvement in the total number of trips and total distance covered by taxis using the proposed method. This work would be helpful for taxi service providers in decision making such as to find the optimal number of taxis to be operated, optimal placement of taxis, and to generate the recommendations for taxi drivers for pickup of every next customer at earliest.

References

1. Yu, H., Li, Z., Zhang, G., Liu, P., Wang, J.: Extracting and predicting taxi hotspots in spatiotemporal dimensions using conditional generative adversarial neural networks. IEEE Trans. Veh. Technol. **69**(4), 3680–3692 (2020)
2. Li, M., He, D., Zhou, X.: Efficient kNN search with occupation in large-scale on-demand ride-hailing. In: Borovica-Gajic, R., Qi, J., Wang, W. (eds.) ADC 2020. LNCS, vol. 12008, pp. 29–41. Springer, Cham (2020). https://doi.org/10.1007/978-3-030-39469-1_3
3. Zhou, D., Hong, R., Xia, J.: Identification of taxi pick-up and drop-off hotspots using the density-based spatial clustering method. In: CICTP 2017: Transportation Reform and Change-Equity, Inclusiveness, Sharing, and Innovation, pp. 196–204. American Society of Civil Engineers, Reston (2017)
4. Chang, H.W., Tai, Y.C., Hsu, J.Y.J.: Context-aware taxi demand hotspots prediction. Int. J. Bus. Intell. Data Min. **5**(1), 3–18 (2010)
5. Fagin, R., Kumar, R., Sivakumar, D.: Efficient similarity search and classification via rank aggregation. In: SIGMOD, pp. 301–312 (2003)
6. Dwork, C., Kumar, R., Naor, M., Sivakumar, D.: Rank aggregation methods for the web. In: WWW, pp. 613–622 (2001)
7. Mamoulis, N., Cheng, K.H., Yiu, M.L., Cheung, D.W.: Efficient aggregation of ranked inputs. In: ICDE, pp. 72–84 (2006)
8. Ailon, N., Charikar, M., Newman, A.: Aggregating inconsistent information: ranking and clustering. J. ACM **55**(5), 23 (2008)
9. Shekhar, S., Feiner, S.K., Aref, W.G.: Spatial computing. Commun. ACM **59**(1), 72–81 (2016)
10. Tao, Y., Hristidis, V., Papadias, D., Papakonstantinou, Y.: Branch-and-bound processing of ranked queries. Inf. Syst. **32**(3), 424–445 (2007)
11. Li, M., Bao, Z., Sellis, T., Yan, S.: Visualization-aided exploration of the real estate data. In: Cheema, M.A., Zhang, W., Chang, L. (eds.) ADC 2016. LNCS, vol. 9877, pp. 435–439. Springer, Cham (2016). https://doi.org/10.1007/978-3-319-46922-5_34
12. Mouratidis, K., Bakiras, S., Papadias, D.: Continuous monitoring of top-k queries over sliding windows. In: SIGMOD, pp. 635–646 (2006)
13. Cormode, G., Hadjieleftheriou, M.: Finding frequent items in data streams. PVLDB **1**(2), 1530–1541 (2008)
14. Papapetrou, O., Garofalakis, M., Deligiannakis, A.: Sketch-based querying of distributed sliding-window data streams. PVLDB **5**(10), 992–1003 (2012)
15. Bohm, C., Ooi, B.C., Plant, C., Yan, Y.: Efficiently processing continuous k-NN queries on data streams. In: ICDE, pp. 156–165 (2007)
16. Korn, F., Muthukrishnan, S., Srivastava, D.: Reverse nearest neighbor aggregates over data streams. In: PVLDB, pp. 814–825 (2002)
17. Li, C., Gu, Y., Qi, J., Yu, G., Zhang, R., Yi, W.: Processing moving KNN queries using influential neighbor sets. PVLDB **8**(2), 113–124 (2014)
18. Cheema, M., Zhang, W., Lin, X., Zhang, Y., Li, X.: Continuous reverse k nearest neighbors queries in Euclidean space and in spatial networks. VLDB J. **21**(1), 69–95 (2012). https://doi.org/10.1007/s00778-011-0235-9
19. Khetarpaul, S., Gupta, S.K., Malhotra, S., Subramaniam, L.V.: Bus arrival time prediction using a modified amalgamation of fuzzy clustering and neural network on spatio-temporal data. In: Sharaf, M.A., Cheema, M.A., Qi, J. (eds.) ADC 2015. LNCS, vol. 9093, pp. 142–154. Springer, Cham (2015). https://doi.org/10.1007/978-3-319-19548-3_12

20. Xia, T., Zhang, D., Kanoulas, E., Du, Y.: On computing top-t most influential spatial sites. In: PVLDB, pp. 946–957 (2005)
21. Li, C.-L., Wang, E.T., Huang, G.-J., Chen, A.L.P.: Top-n query processing in spatial databases considering bi-chromatic reverse k-nearest neighbors. Inf. Syst. **42**, 123–138 (2014)
22. Koh, J.-L., Lin, C.-Y., Chen, A.P.: Finding k most favorite products based on reverse top-t queries. PVLDB **23**(4), 541–564 (2014). https://doi.org/10.1007/s00778-013-0336-8
23. Vlachou, A., Doulkeridis, C., Nørvåg, K., Kotidis, Y.: Identifying the most influential data objects with reverse top-k queries. PVLDB **3**(1–2), 364–372 (2010)
24. Wong, R.C.-W., Özsu, M.T., Yu, P.S., Fu, A.W.-C., Liu, L.: Efficient method for maximizing bichromatic reverse nearest neighbor. PVLDB **2**(1), 1126–1137 (2009)
25. Gkorgkas, O., Vlachou, A., Doulkeridis, C., Nørvåg, K.: Discovering influential data objects over time. In: Nascimento, M.A., et al. (eds.) SSTD 2013. LNCS, vol. 8098, pp. 110–127. Springer, Heidelberg (2013). https://doi.org/10.1007/978-3-642-40235-7_7
26. Choudhury, F.M., Bao, Z., Culpepper, J.S., Sellis, T.: Monitoring the top-m aggregation in a sliding window of spatial queries (2016)
27. Sampathkumar, E., Walikar, H.B.: Connected domination number of a graph. J. Math. Phys. **13**, 1–7 (1979)
28. Pang, C., Zhang, R., Zhang, Q., Wang, J.: Dominating sets in directed graphs. Inf. Sci. **180**(19), 3647–3652 (2010)
29. He, H., Zhu, Z., Makinen, E.: A neural network model to minimize the connected dominating set for self-configuration of wireless sensor networks. IEEE Trans. Neural Netw. **20**(6), 973–982 (2009)
30. http://users.diag.uniroma1.it/challenge9/download.shtml
31. https://www1.nyc.gov/site/tlc/about/tlc-trip-record-data.page

Adaptive Fault Diagnosis for Data Replication Systems

Chee Keong Wee[1]([✉]) and Nathan Wee[2]([✉])

[1] Database Team, Energy Queensland Limited, Townsville, QLD, Australia
ck@outlook.com
[2] Faculty of Science, University of Queensland, Brisbane, QLD, Australia
nathan.wee@uq.net.au

Abstract. Data replication among multiple IT systems is ubiquitous among large organizations and keeping them running is a critical success factor for their IT departments. When services are disrupted, IT administrators must be able to find the faults and rectify them quickly. Due to the scale and complexity of the data replication environment, the fault diagnostic effort is both tedious and laborious. This paper proposes an approach to fault diagnosis of the data replication software through deep reinforcement learning. Empirical results show that the new method can identify and deduce the software faults quickly with high accuracy.

1 Introduction

Data replication is one of the essential IT services in a large organization where data is distributed and shared to service various business needs, and these replicating services are conducted by numerous commercial software which had been built for this purpose at very low latency. Their uptime and service are critical to the business' functions, so ensuring that they are operating at an optimum level is important [1]. However, software in a complex IT environment will face operational issues and faults that can be attributed to various reasons such as issues arise from underlying operating systems, network connectivity, permission, and many other causes [2]. The amount of effort to troubleshoot and resolve any IT fault consume the bulk of any IT administrator's work time and it varied from a short period if the fault is easy to fix to long-duration where the fault may require software vendors to develop bug fixes plus the cycle of product acceptance testing. The extent of this will exacerbate when multiple faults are occurring concurrently and there is a limited number of IT administrators available to handle the job. The IT administrators' manual hands-on effort is not well known for scalability for a larger number of IT systems, coverage in the period of support throughout the period on 24×365. In addition to that, they are prone to fatigue, human errors, and slowness. The capability of any IT administrator is also limited to their skills and experience including other social or human conditions too. As the data replicating environment comprise of a multitude of software and hardware technology, any single IT administrator will find it difficult to maintain a certain expert level of expertise across multiple software domains [3].

© Springer Nature Switzerland AG 2021
M. Qiao et al. (Eds.): ADC 2021, LNCS 12610, pp. 125–138, 2021.
https://doi.org/10.1007/978-3-030-69377-0_11

We propose a novel method to conduct fault detection and diagnosis for the data replicating setup using deep reinforcement learning. To our best knowledge, there is no prior research on applying Machine learning for fault diagnostics in the field of data replication for databases. The computational search space to align the combination of information from the Data Replicating Environment (DRE) versus the possible root cause is high and difficult. For any single fault detected, there may be other various forms of causes. Techniques like decision trees, Bayesian networks, or statistical methods are commonly used in the industry to meet this need. The diagnostic models are customized to the solution that they are intended for, and they are not readily modifiable to fit in new changes nor interoperable with other software domains. We want our new solution to be general-purpose enough to adapt to any software domain and with greater flexibility and ability to expand its knowledgebase of diagnosing faults with ease.

2   Literature Review and Background

A common implementation of fault diagnosis with machine learning involves the acquisition of signals or data, the perform feature extraction, or information infusion before using machine learning models to perform pattern recognition to derive fault diagnosis [4]. Another approach is the use of a convolutional neural network for better accuracy in fault diagnosis for industrial's permanent magnet synchronous motors in conjunction with the analysis of generalized frequency response function for [5].

Deep neural networks are used prevalently in both academia and industry for the system's fault diagnosis across different domains. But there are some shortcomings that researchers have identified in the use of NN for this approach and they are; 1) the complexity of mapping relationship between data and outcome of faults for complex, non-linear systems, 2) the availability of labelled data and high quality extracted features for the model training, 3) the configuration of the neural networks models need frequent retraining, reconfiguration and optimization to keep them relevant [6].

The authors use Reinforcement Learning (RL) to diagnose the ball bearing faults in a motor, by acquiring the signals from monitoring meters set against various components within the motor and use Neural Network (NN) to predict the outcome [7]. The outcome is compared against a set of labelled data that has been predetermined to gauge the quality of their predictions. A q-table is built between the detected motor's conditions and the outcome of the predictions during training, which is used by the agent during its knowledge exploitation phase. Another research [8] followed a similar approach with a signal band filter in its fault determination criteria on signals data gained from rotary machinery motors. These are initial research conducted in using deep reinforcement learning for motor machine fault diagnosis and we observed that the application has several features; it is an enclosed system with minimum or no direct interaction with external entities. All the data input are signals from monitoring meters set around the motor, and the data acquired are explicit homogeneous. There is a set of prelabelled data that the RL's NN can refer to. However, in a complex IT system environment, it is an open system with many interfacing components and the data received for fault diagnosis are heterogeneous in data type, categories, frequency of occurrence and status usage. It requires an additional set of data manipulation and a new design before RL can be used for this intent. The authors in [9] used RL for fault diagnosis on a virtualized network in the cloud service, acquiring inputs from a list of subsystems within the Virtual Machine (VM) networks and map to state-attribute vectors. It used an external entity to validate the attributes in response to the state. The attributes are regarded as the diagnosed faults and are manually mapped to external actions to remediate them. This gives rise to the inspiration of a similar approach for this paper, but the main difference is in the setup of the diagnosis validation module where we use a script-based approach as compared to the manual mode in this research [9].

The current common method of implementing fault diagnosis for complex IT systems for both academia and industry is to use machine learning models such as Random Forests or Bayesian Network [10]. Both require well-designed models that are specifically tailored to the intended IT systems where the fault detection and diagnosis procedures need to be performed. The premise for the design of such complex and well-defined Fault Detection and Diagnosis (FDD) model has complete knowledge of every sub-system, components, relationship, and operations including data exchange in the IT system. The limitation with this approach is that every implementation of these complex IT systems is not generic and are tailored to specific business IT requirement [11]. So, having a rigid and well-defined FDD agent will not have the adaptiveness nor flexibility to meet the range of different system setup. It will require numerous customization which is time-consuming and laborious.

What is required here is a new approach where the FDD model can be made general-purpose enough to suit any combination of software for the IT systems; be it database, web application, firewall, or network. It should minimize unnecessary steps of detailed check procedures and able to deduce the diagnosis quickly simply by looking at the symptoms and refer to its knowledge just an experienced IT administrator. It should be flexible to extend or correct its existing model to cover any new alteration that occurs in the IT system's environment. In another word, we relate the new FDD model as a new mechanic apprentice that need to learn on the job to perform the checks and deduce the faults from the gathered information under the guidance of his supervisor. We expect it to learn in both detecting and diagnosing adaptively, starting from an early stage where it will do extensive checks on every aspect of the IT system, but once it reaches a certain level of maturity, it should be able to determine from its expert knowledge that the certain symptoms or events exhibited in the IT system can be related to certain sub-domain of the system's setup with great confidence, similar to the skill difference between an inexperienced and an expert IT administrator.

3   Adaptive Fault Diagnosis (FD) Module Design

Deep Reinforcement Learning (DRL) [12] is used in performing the fault diagnosis against the Data Replication Environment (DRE) with the objective of an intelligent system that can emulate the learning and work process that is similar to a junior IT administrator in managing a system related problem. Figure 1 shows an overview of the FDD model. As an individual that is receiving on-the-job training, it will learn initially to analyze every aspect of the DRE and gather all the related information. It also starts with a little or no prior knowledge on the relationship between information and diagnostic conclusions, it interacts with its supervisor or someone with expert domain knowledge, receiving guidance and information to derive the diagnosis. This iteration carries on and the FDD will build up its knowledge slowly, and when it has sufficient knowhows on the environment, it can derive an accurate deduction of the fault just by looking at the environment's symptoms without going through all the unnecessary and unrelated system checks [12]. The justification for using DRL for supporting the fault diagnosis is mentioned in problem definition, while these routines can be easily performed by hardcoded, well-defined models, the scope for these groups of software fault diagnostic system cannot be fixed. They must be flexible to cater to different fault scenarios and grow to cover other forms of the software system that can be rolled under the FD's management. A series of data collection points are established to ingest and process the information from the DRE before sending it to the FDR's DRL unit [13]. Once the agent has determined that there is a need to investigate, it will launch a series of queries to acquire more details from the environment for its fault analysis and diagnostic routine until it can reach a point where it can either deduce the root cause or listed it as an unknown error, which triggers another routine to notify the IT administrator for assistance and input. The FDD model can be split into 3 modules: Information Acquisition (IA), Diagnostic Reinforcement Learning (DRL), and System Diagnostic (SD) as shown in Fig. 2 [12].

3.1 Information Acquisition (IA) Module

The availability of timely and accurate information from various software subsystems of the DRE is important to the diagnostic analysis process and they come in three forms: logs, metrics, and events. All the DRE's software; Oracle Database (DB) [14], Shareplex [15], and Operating System (OS) [16], produce information about their states constantly and proactively into log files under the software's respective product directories. They are available in a well-defined format and provide enough information to support a system diagnosis effort. For the metrics part, these are system or software statistics that can only be obtained through explicit command queries via OS' shells or their respective utility tools. The third form is the events that logically describe the experiences from the users or another depending system while interacting with the DRE. It is a brief description of encountered service's anomalies for the FDD can refer to for investigation. An event such as login failure, slow replication, data not found as some examples. Once the inputs have been processed with all the mandatory details extracted out, they are used to represent the system environment's state to the DRL and as input into both the SD modules.

3.2 Diagnostic Reinforcement Learning (DRL) for FD Module

The FDD module uses the Actor-Critic Deep Reinforcement Learning algorithm (DRL) [17, 18]. Referring to Fig. 2, in this setup, the DRL's Actor is performing as a function approximator that tries to predict the best action for a given state, and in this case, the best diagnosis. The DRL's Critic also takes in the DRE' state-input plus the Actor's action, join them and output the action's maximum future reward, Q-value, for the given state-action. The Critic uses the SD module to validate and score the Actor's action. There are 3 phases of learnings for the DRL as shown in Fig. 3 [18].

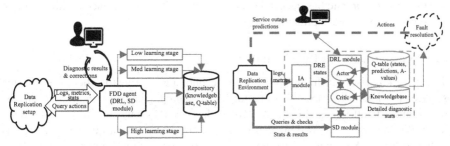

Fig. 1. Adaptive faults diagnosis overview

Fig. 2. Faults diagnosis agent's architecture and workflow

1) Early learning phase; where the agent has little a-prior knowledge about the DRE at the start, so it needs to perform exploration by interacting with it through trial-and-error. It passes the DRE's states into the SD module as symptoms and the SD module runs through all the diagnostics routine against the DRE's software to gather information about their attributes and service status. The SD module will then be processed and summarised the acquired information to formulate a service diagnosis matrix as shown in Eq. (1) [18].

$$DRE's\ diagnosed\ service\ matrix = \begin{bmatrix} sdbs_1\ srpl_1\ snet_1\ sose_1 \\ sdbs_2\ srpl_2\ snet_2\ sose_2 \\ \cdots\quad \cdots\quad \cdots\quad \cdots \\ sdbs_n\ srpl_n\ snet_n\ sose_n \end{bmatrix} \tag{1}$$

2) Middle learning phase; after it has gathered enough knowledge about the DRE, the DRL's Actor learns to predict the best action-diagnosis against the environment's symptoms-states using its neural network which has been trained by using the knowledgebase gathered from the earlier learning phase as its minibatch. There is a high chance that the NN will predict incorrectly, so in such an event, the RL's Critic will run the validation process through the SD module which corrects and assign the Q-value to the state-action pairs and then store in the Q-table, as well as updating the knowledgebase. The gradual built-up of the knowledgebase will improve DRL's NN prediction accuracy [18].

3) High learning phase. By this stage, the DRL agent would have learned all the states - symptoms that may associate with the faults in the DRE and can predict the best actions-diagnosis with high accuracy. This is regarded as the exploitation of the DRL's rich build-up of knowledge where it can provide a very quick turnaround time in identifying the faults' matrix without performing excessive checks or validation through the SD module. But during this period, the agent also performs a probability-based decision between exploitation versus exploration; Exploitation where the DRL decided to refer to its knowledgebase to respond the best action for the DRE's state, Exploration where DRL's decide to run all the detailed checks through SD module and get the diagnosis instead of relying on the NN's prediction. At the start of the learning cycle, the probability for exploration will be high at the low learning phase but this diminishes over time when it reaches the high learning phase, where

the exploration rate has decayed over iterations and the preference is shifting more toward knowledge exploitation [18].

3.3 System Diagnostic (SD) Module

The DRE comprise of different software and technology working together to provide the service [19]. Each software and technology have a unique list of configuration, checks, operations, and attributes. So, the SD Module has several groups of check routines that target this software, and within each group are sub-routines that query specific areas in the software like privileges, permission, process status, usage statistics, and others. Referring to Fig. 2, the DRL agent gives instructions to the SD module to do the checks against DRE's environment, ranging from comprehensive top-down checks or a few selective ones. This is on par with the analogy of junior workers that need to perform every check just to make sure, or a senior worker who can deduce roughly which exact area that has to be verified before deducing the root cause. The SD module then performs the detailed checks by running a long list of command queries and scripts against the DRE's software. Some of the details collected are the 1) states of their system processes, 2) space availability of directories in which the systems' binary files reside on and their information are processed, 3) current privileges of the system's process, files, accounts that they operate from, 4) details in their configurations and parameters that they are using, operating or initialize from, 5) network connectivity that is required for their operation, 6) statistics of specific operations like process backlogs, connectivity delays, abnormal system values. Others contain a summary of software-wide statistics which range in the thousands. The result is then consolidated as shown in Table 1 and sent back to the RL agent. It is a matrix that presents the multiple sub-area under the DRE across different software about their functional status from a high-level perspective. Further details can be made available from the diagnostic module upon request, but the vast amount of details will be too overwhelming for its administrators to go through. The following is a tabulation of the output which each command performing the specific information extraction from the various software.

The SD provides its diagnosed results of the DRE's software status on the participating server hosts, n, at their service group level instead of the technical attributes. This is to give an overview of the DRE software's availability from a general administrative perspective; taking into consideration their 1) process availability, 2) filesystem's attributes and permission, 3) responsiveness to administrative interaction, 4) communication functionality, 5) data transfer and input-output capability plus 6) software's function and operation status. The vast specific software details can be made available and they will be connected to the future Fault Resolution agent. The four diagnosed service groups are as followed and in the matrix in Eq. (1); Database service, $sdbs_n$, Shareplex replication service, $srpl_n$, Network and communication services, $snet_{n,}$ supporting the OS environment, $sose_n$.

4 Data Replication Environment (DRE)'s State Representation

For the DRE [18], it is hard to define its state due to its complex multi-tier software setup and the characteristics of the IT applications under its service. A direct method is needed to identify a state in a database without time properties. Each software's operation information is mined continuously for anomalies and errors. We propose the use of a matrix to capture a list of the events and processes' status of the DRE's software across multiple sources and target instances, n. Therefore, the two sections in the state's matrix contain both information from both their logs and process status. For the logs, the attributes are a numerical representation of the encountered error messages in their respective logs, which are concatenated to 10 characters long and hashed using Secure Hash Algorithm 1 (SHA1). The following is the list of the software's logs location and their respective variables assigned.

1. Oracle database's alert logs with the prefix of ORA-XXX, files exist in the location; $ORACLE_BASE/diag/rdbms/DB1/trace/alert_DB1.log, as $oralog_n$.
2. Shareplex replication's event_logs with the initial string of "Error", files available in the location at; $VARDIR/log/event_log, as $splxlog_n$.
3. Network-related Listener's logs with the prefix of LSNR-XXX, available in $ORACLE_HOME/diag/network/log/.log as $nwlog_n$.
4. OS's error with the string, err, in /var/log/syslog, as $oslog_n$.

For the process's status, the status shows the presence of the DRE's software main processes in the VM host's background as well as the reachability of remote VM from the current VM. The representations are 1) Oracle DB's primary process, smon, as $orastat_n$. 2) Shareplex replication's main process, sp_cop, as $splxstat_n$. 3) Oracle's listener's processes and network, lsnrctl, as $nwstat_n$. 4) Ping status from both UNIX nodes to one another, as $osstat_n$. The services under the different software are represented as; 1) Oracle DB's as $orasvc_n$. 2) Shareplex replication as $splxsvc_n$. 3) Oracle's listener and network, as $nwsvc_n$. 4) Operating system and host's, as $ossvc_n$.

Therefore, the final matrix to represent the DRE's state in Eq. (2).

$$DRE's\ state = \begin{bmatrix} oralog_1 \\ oralog_2 \\ splxlog_1 \\ splxlog_2 \\ nwlog_1 \\ nwlog_2 \\ oslog_1 \\ oslog_2 \end{bmatrix} \begin{bmatrix} orastat_1 \\ orastat_2 \\ splxstat_1 \\ splxstat_2 \\ nwstat_1 \\ nwstat_2 \\ osstat_1 \\ osstat_2 \end{bmatrix} \begin{bmatrix} orasvc_1 \\ orasvc_2 \\ splxsvc_1 \\ splxsvc_2 \\ nwsvc_1 \\ nwsvc_2 \\ ossvc_1 \\ ossvc_2 \end{bmatrix} \quad (2)$$

Table 1 describes the specific software validation and checks that need to be performed to acquire the DRE' collective status together with the associated details that depict their respective software components including the checks are performed against them. Each of the software is checked by different OS scripts which have encapsulated commands to interrogate them on their respective service groups of logs, processes, and services. For various software logs check, the scripts are check_alert_log_err.sh, check_event_log_err.sh, check_os_log_err.sh on OracleDB with listener, Shareplex and OS. As for all the DRE's software processes checks, check_all_processes.sh will handle

Table 1. Subsets of memory and logs checks

DRE's software	Service Group checks	Software attribute	Detail checks/description
OracleDB	Process check	Process' stats	DBs' memory process in the OS
	Process check	Operation's stats	DB's mode of operation
	Service check	Tablespace's stats	Tablespaces have enough space on DBs
	....		
Shareplex	Service check	Parameter setting	parameters are valid in Shareplex instances
	Process check Service check	Process's Status and operation	Shareplex's memory process in the OS
	....		
Network	Process check	Listener Process stats	Oracle's listener process on both nodes
	Process check, Service check	Listener.ora availability, and stats	Listeners' availability for service on both nodes
	Process check, Service check	Listener's stats – error or available	Listeners' operations are valid and not in error
	....		
OS	Service check	Disk space	Free space availability on OS for both nodes
	Service check	Primary conf files	Validate /etc/passwd, /etc/shadow, /etc/hosts, /etc/group files
	Process check	Network card operation	Network card status and availability
	....		

this. The last group check is done by check_all_services.sh which validates their specific services.

5   DRE's Action of Diagnostic Prediction

The DRE's state information from the previous section are the summarised raw input which the FDR takes in, and part of its diagnostic routine is to show its ability to predict or estimate the possible faults with the DRE's software, much like an experienced mechanic that can pinpoint the fault with a car based on the symptoms described by the owner [18]. Part of the outcome of the FDR is to produce the diagnostics report that shows the status of the DRE's operation at a high service level which indicates the software's respective sub-group and level of errors it has, in respect to the DRE's environment state. The outcome is a series of tuples that signify the status or condition of the software group and their sub-group services in the arrangement of <software_typ> and <software_sub-service_grp>. Their statuses are derived from a custom-built script which contains a list of OS commands that extract and aggregate all the statistics from the various DRE software into their respective sub-system service groups, as a mean to show the service outage based on the state's matrix from the environment in the previous chapter. The process of showing the service-level exceptions will be later handled by the DRL's NN.

DRE's service level diagnosis = {dba, dbb, dbc, dbd, spa, spb, spc, spd, spe, spf, nwa, nwb, nwc, osa, osb, osc, osd}.

Where,

1. for OracleDB, dba = DB's memory process, dbb = DB's Status, dbc = DB's Account security, dbd = DB's storage space.
2. for Shareplex, spa = Splx's main processes, spb = splx's console availability, spc = splx's queues operation, spd = splx's configuration validity, spe = splx's queues' backlogs, spf = Splx's DB accessibility.
3. for the networks, nwa = Network connectivity of Databases' listeners, nwb = Splx's network connectivity, nwc = VM hosts interconnectivities.

4. for the OS, osa = hosts' OS unix account status, *osb* = hosts' file storage space, *osc* = hosts' network card status, *osd* = hosts' resource availability.

6 Empirical Analysis

This section describes the tests conducted for the FD module. The purpose of experiments is to determine the effectiveness of the proposed FD method in producing the best diagnosis for the DRE under simulated faults situation [18]. Before each experiment's iteration, the testing environment DRE's services are restored to the baseline where all the DRE's services are functioning normally. Not all errors introduced can result in a service's disruption. The goal is to ascertain the diagnosis on those faults that can disrupt the services and less toward those that are either too minor or ineffective to cause major issues to the replication services. However, the test scope is limited to faults that are recoverable and not on catastrophic failure, which is irrecoverable and can only be solved by an entire system rebuild.

6.1 The Experimental Set-Up

The experiments are run on two Virtual Machines running on Linux OS and both have Oracle DB and Shareplex installed on them. Each VM has 4 GB of RAM with 100 GB of hard disk storage. The version of the Oracle software is 12 Enterprise edition and the 9.1 for the Shareplex. The network protocol that both VMs use is TCPIP. For the DBs, the simulated faults will impact Oracle's primary memory process such as SMON and PMON. Any failure of either one of these processes will cause the DB service to stop. The script will do a root level kill to simulate the DB outage and a start-up command is required via DB's admin level is required in restoring it. The fault-inducing and correcting scripts will modify the user account status to be in open or locked mode. The Shareplex also require a user account to have a list of DB level privilege to function, so some scripts simulate the absence and presence of these privileges from the accounts. Likewise, for the schema objects that the user account owns and access; the Shareplex created a list of DB objects under the user account during installation and it continues to use them for its operation. Should there be any changes to their accessibility to the user account of the validity of the object, it will cause Shareplex to malfunction. Scripts are written to simulate this error too. Another factor to note is the availability of free space within DB for the Shareplex to operate on. If there is insufficient space, then Shareplex will not be able to write data into the DB and that results in the suspension of its service. Some scripts constrict and free up the storage space. For the Shareplex's fault simulation, it follows a similar pattern as the DB, with the focus on their instance's primary processes that run on the OS. Their service disruption and restoration are done by scripts that execute system-level commands against their console.

As for the network inter-connectivity, there are two main areas in which the fault can be induced for this setup; 1) the connection via the TCPIP protocol at the OS level between the two VM hosts and, 2) the ability of the software's client to connect to the current and remote DBs through the oracle's network grid which comprises of listener services, OCI library, and oracle-related network files setup. The scripts that perform the opposing functions of faults induction and restoration target the network card's status, the listener process availability and status, the presence and validity of the network configuration files, as well as the OS' network files under the /etc. folders.

For the OS, the emphasis here is on 1) the Unix user accounts that Oracle and Shareplex need to use throughout their services, 2) the availability of free space on the disk partitions that their home and operational directories are installed on, 3) the resource availability in the OS which both Oracle and Shareplex can operate under and 3) status of the network card. For each of the software's core functionalities, two of its attributes will be assessed and a metric is associated with it which measures its service's normality. A value of 0 indicates a normal state whereas >0 indicates an abnormality. Table 2 lists all the software components and the respective commands that can simulate and restore their faults.

However, the test does not include malicious or terminal faults to the software if they are either irreversible or require a substantial amount of effort to restore them. Examples of such faults are the corruption or deletion of the software's binaries or libraries, deletion of DB's repository, file-based data store and erasure of OS' disk mount-point. The neural network that the RL used for its rewards-action prediction is made up of 3 hidden layers of 30 nodes. It is trained with data in 50 batches and 500 epochs. Different configurations and combinations of neural networks have been tested, and this setup was selected based on the better results with the least fluctuations.

6.2 True Negative Test Results

Besides the data are obtained from the faults inducing scripts in the previous section, another group of scripts has been created to induce software faults that have no impact on their DRE's software functionalities and services. This is to form the set of true negative data to support and enrich the dataset for the NN's training so that the NN can be competent enough to recognize and differentiate the environment's state data that can cause service disruption or not.

For the script to induce this group of faults, research has been made across the DRE's software to identify those faults that have a high chance of occurring but they don't have a direct consequential effect that can either disrupt the entire software's stability or create outage on the DRE's functionalities. This is verified by the SD module which confirms the presence of any service disruption. For this group, the service disruption matrix values should all be zero. Once these faults are induced, the software will capture their exceptions and events in their event or trace logs, which in turn are detected by the FD module.

6.3 Evaluation Criteria and Benchmarking

This section describes how the FD module is evaluated and the criteria used in its assessment. The faults statistics cover the four main DRE's software; Database, Replication, Network, operating system, and service level are represented by a vector with each element representing the service. And within each element is a scalar value from 0 to 1, values that are >0 indicate the faults' severity whereas 0 is when every component is

Table 2. Faults induction and restoration on DRE software's component services (service status flag: 0 – good, 1 – faults)

Software	Component/services	Target for faults	Fault inducing action	Service restoring action
Databases	Memory process	PMON, SMON processes availability	Kill off PMON process Kill off SMON process	Start oracle instance (which start both PMON and SMON)
	Status	DB operational and service status	Shutdown and start in mount mode	Open DB for use
	Account security	DB's System and splx accounts' status Splx has quota on splx tablespace	Lock up system and splx DB account Splx user has no quota on tablespace to write	Unlock system and splx DB user account Splx has quota to write on tablespace
	DB storage space	Amount of free space in system and splx tablespaces	Shrink tablespace to 100% full	Increase tablespace space to have 20% of free space
Shareplex replication	Mmain processes	Shareplex main processes availability Sp_cop, Capture, Read, Exp, Imp, Post processes	Kill off individual processes	Restart sp_cop to resume all processes
	Queues' operation	Capture, Export, Import, Post and Read's queues	Stop the queues' operations	Start the queues' operations
	DB accessibility	DB connection using splx Unix account from current and opposite VM hosts	Lock DB user account	Unlock DB user account
Network connectivity	Oracle listeners	Source & target Listeners Source & target host connect to target DB via sqlplus	Stop the listener process to stop user from connecting to on-site DBs	Start the listener process to allow user to connect to on-site DBs
	Oracle network files	Essential files availability; tnsnames.ora, listener.ora	Delete off network files	Restore network files
	VM hosts	Each VM host can reach the opposite node	Disable sshd service	Enable sshd service
Host OS	Unix account status	splx and oracle's Unix accounts	Lock the Unix user accounts	Unlock the Unix user accounts
	Essential OS system files	Essential Unix files like /etc./hosts	Delete the /etc./hosts file	Restore /etc./hosts file
	Network card status	Network service on enps03network cards on both hosts	Disable network card	Enable network card

operating normally. This forms the basis for the primary evaluation criteria. The statistical differences among fault diagnosis of DRE's states can indicate the progress of the DRE's overall service of whether they are improving or degrading. Each diagnosis is correlated to the detailed diagnostic statistics that were generated by the FD module which will be vital for the next module of fault resolution.

6.4   Test Results

This section described the results obtained from the FD module after it completed the training and subjected to the evaluation test processes. By this stage, the FD module has been trained thoroughly and it is regarded to be equivalent to achieving the expert level of fault diagnostic capability. The minimum expectation of its prediction accuracy internally is expected to reach 85% accuracy and more. A sample of the DRE's states, including both the predicted and actual service outage results, are shown in Table 3; 1) The DRE state data are derived from the information gathered against the DRE's software components from their logs, internal system statistics, and monitoring after simulating fault are induced. 2) The FD module predicted the service outage results after it received the DRE input based on its learned NN. 3) The SD module produced the real detailed results by running a list of diagnostic routines against the DRE environment to derive and aggregate the actual statistics. 4) The classification of the outage results is derived by comparing the sum of the predicted results' values against the actual service outage results. 5) The MASE score is calculated based on the difference in the vectors' values between the predicted and actual results.

Table 3. Results of service outage prediction & scores against DRE's state

DRE State	Service outage Predicted	Service outage Actual with rounding	Classes	MASE
[0,64655058,76223968,0,0,0,0,64351381] [1,0,1,0,0,0,0,0] [1,0,0,0,0,0,0,0]	[[6,1,0,1,3,1],[2,2,1,1,0,0],[1,0,0,0,0,0],[0,0,0, 0,0],[0,0,0,0,0,0]]	[[6,1,0,1,3,1],[2,2,1,1,0,0],[1,0,0,0,0,0],[0,0 ,0,0,0],[0,0,0,0,0,0]]	TP	0.6
[46968001,0,0,0,0,0,0,64351381] [0,1,0,0,0,0,0,0] [0,1,0,0,0,0,0,0]	[[0,0,0,1,0,2],[2,2,2,1,0,0],[1,0,0,0,0,0],[0,0,0, 0,0],[0,0,0,0,0,0]]	[[0,0,0,1,0,2],[2,2,2,1,0,0],[1,0,0,0,0,0],[0,0 ,0,0,0],[0,0,0,0,0,0]]	TP	0.3
[46968001,0,0,0,0,0,0,64351381] [0,1,0,0,0,0,0,0] [0,1,0,0,0,0,0,0]	[[0,0,0,1,0,2],[2,2,2,1,0,0],[1,0,0,0,0,0],[0,0,0, 0,0],[0,0,0,0,0,0]]	[[0,0,0,1,0,2],[2,2,2,1,0,0],[1,0,0,0,0,0],[0,0 ,0,0,0],[0,0,0,0,0,0]]	TP	0.2
[46968001,0,0,0,0,0,0,0] [0,0,1,0,0,0,0,0] [0,0,0,0,0,0,0,0]	[[6,1,0,1,3,0],[0,0,0,0,0,0],[0,0,0,0,0,0],[0,0,0, 0,0],[0,0,0,0,0,0]]	[[6,1,0,1,3,0],[0,0,0,0,0,0],[0,0,0,0,0,0],[0,0 ,0,0,0],[0,0,0,0,0,0]]	TP	0.1
[46968001,0,0,0,0,0,0,64351381] [0,1,0,0,0,0,0,0] [0,1,0,0,0,0,0,0]	[[0,0,0,1,0,2],[2,2,2,1,0,0],[1,0,0,0,0,0],[0,0,0, 0,0],[0,0,0,0,0,0]]	[[0,0,0,1,0,2],[2,2,2,1,0,0],[1,0,0,0,0,0],[0,0 ,0,0,0],[0,0,0,0,0,0]]	TP	0.6
...	...	...	...	...

6.5   Service Outage Classification Results

The test is conducted with a list fault inducing scripts with 80 entries. 30 of them have a direct effect on the software's functionalities which impact the DRE's software services, and 50 of them do not. It is expected that the FD module can predict accurately for both groups. The results are split into qualitative and quantitative groups. Table 4 is the tabulation of the prediction's result classes in a confusion matrix. The results showed that the SD module can predict the group of service outage to the information received from the DRE's environment. While it has high capability in recognizing most of the induced faults that can affect the DRE's software functionalities, it fair less when it comes to the detection of those in the other groups. Based on the result, the FD's sensitivity is 0.355, specificity is 0.645, precision is 0.871. The SD module has shown to be accurate

Table 4. Confusion matrix of the classification of the service outage's prediction

N = 80	Predicted: yes	Predicted: no
Actual: yes	27(TP)	4(FN)
Actual: no	1(FP)	49(TN)

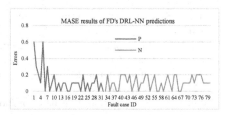

Fig. 3. MASE score of True and positive predicted results

enough that its prediction can produce the correct category of service outage for the given environment state's data input. It has the competency to differentiate if the inputs are related to DRE's service functionalities.

6.6 Service Outage's Prediction Accuracy

For this test, The SD module forms the baseline in which the FD's predictions are measured against. Each value in the service outage results produced by both the FD and SD is calculated using the Mean Square Error approach, and they are summed up to form the total overall degree of accuracy for the SD. The results are shown in the chart in Fig. 3. Based on the results, the accuracy is below the mark of 0.3 and below except for one entry that scored 0.6. This can be since this is a DRL based Fault diagnosis that learns adaptively with the environment. While it is experienced to recognize the fault scenario that it was had trained for. However, for new and unfamiliar ones, it has some deviations. One possible solution is to enable more iterations of exposure for the SD module's DRL to learn more about the true positive and negative of DRE's scenario. However, the list of potential faults that can affect DRE is controlled and limited, the next possible solution is to expose the NN to the more true-negative class scenario which does not impact the DRE. This has more potential to be generated in greater volumes and can assist in enriching NN's training dataset.

7 Conclusion

The FD module has been proven that it be able to produce the outcome of the service outage based on the DRE's state information with good accuracy. It made use of the model-free actor-critic Deep Reinforcement learning to learn against the Data replication setup predict the outage gradually as it interacts with it and learns with the help of the SD module that corrects its prediction. It is adaptive to the DRE and available to configure to support heterogeneous platforms and software without restriction to the data replication architecture.

References

1. Milani, B.A., Navimipour, N.J.: A systematic literature review of the data replication techniques in the cloud environments. Big Data Res. **10**, 1–7 (2017)

2. Tabet, K., et al.: Data replication in cloud systems: a survey. Int. J. Inf. Syst. Soc. Chang. (IJISSC) **8**(3), 17–33 (2017)
3. Iacob, N.: Data replication in distributed environments. Ann.-Econ. Ser. **4**, 193–202 (2010)
4. Chen, X., Feng, Z.J.: Time-frequency space vector modulus analysis of motor current for planetary gearbox fault diagnosis under variable speed conditions. Mech. Syst. Signal Process. **121**, 636–654 (2019)
5. Chen, L., Zhang, Z., Cao, J.: A novel method of combining generalized frequency response function and convolutional neural network for complex system fault diagnosis. PLoS ONE **15**(2), e0228324 (2020)
6. Jia, F., et al.: A neural network constructed by deep learning technique and its application to intelligent fault diagnosis of machines. Neurocomputing **272**, 619–628 (2018)
7. Ding, Y., et al.: Intelligent fault diagnosis for rotating machinery using deep Q-network based health state classification: a deep reinforcement learning approach. Adv. Eng. Inform. **42**, 100977 (2019)
8. Dai, W., et al.: Fault diagnosis of rotating machinery based on deep reinforcement learning and reciprocal of smoothness index. IEEE Sens. J. **20**, 8307–8315 (2020)
9. Xu, T., et al.: Fault diagnosis for the virtualized network in the cloud environment using reinforcement learning. In: 2019 IEEE International Conference on Smart Cloud (SmartCloud). IEEE (2019)
10. Venkatasubramanian, V., et al.: A review of process fault detection and diagnosis: part I: quantitative model-based methods. Comput. Chem. Eng. **27**(3), 293–311 (2003)
11. Venkatasubramanian, V., Chan, K.: A neural network methodology for process fault diagnosis. AIChE J. **35**(12), 1993–2002 (1989)
12. Zhang, D., Lin, Z., Gao, Z.: A novel fault detection with minimizing the noise-signal ratio using reinforcement learning. Sensors **18**(9), 3087 (2018)
13. Zhang, D., Gao, Z.: Reinforcement learning–based fault-tolerant control with application to flux cored wire system. Meas. Control **51**(7–8), 349–359 (2018)
14. Kyte, T., Kuhn, D.: Expert Oracle Database Architecture. Apress, New York (2014)
15. Quest Software: Shareplex for Oracle v9.1.4 (2018)
16. Shotts, W.: The Linux Command Line: A Complete Introduction. No Starch Press, San Francisco (2019)
17. Fujimoto, S., Van Hoof, H., Meger, D.: Addressing function approximation error in actor-critic methods. arXiv preprint arXiv:1802.09477 (2018)
18. Wee, C.K., Nayak, R.: Adaptive database's performance tuning based on reinforcement learning. In: Ohara, K., Bai, Q. (eds.) Knowledge Management and Acquisition for Intelligent Systems. PKAW 2019. Lecture Notes in Computer Science, vol. 11669. Springer, Cham (2019). https://doi.org/10.1007/978-3-030-30639-7_9
19. Wee, C.K., Nayak, R.: Data replication optimization using simulated annealing. In: Le, T., et al. (eds.) Data Mining. AusDM 2019. Communications in Computer and Information Science, vol. 1127. Springer, Singapore (2019). https://doi.org/10.1007/978-981-15-1699-3_18

Entropy-Based Uncertainty Calibration for Generalized Zero-Shot Learning

Zhi Chen[1(✉)], Zi Huang[1], Jingjing Li[2], and Zheng Zhang[3]

[1] The University of Queensland, Brisbane, Australia
uqzhichen@gmail.com, huang@itee.uq.edu.au
[2] University of Electronic Science and Technology of China, Chengdu, China
ijin117@yeah.net
[3] Harbin Institute of Technology, Shenzhen, China
darrenzz219@gmail.com

Abstract. Compared to conventional zero-shot learning (ZSL) where recognising unseen classes is the primary or only aim, the goal of generalized zero-shot learning (GZSL) is to recognise both seen and unseen classes. Most GZSL methods typically learn to synthesise visual representations from semantic information on the unseen classes. However, these types of models are prone to overfitting the seen classes, resulting in distribution overlap between the generated features of the seen and unseen classes. The overlapping region is filled with uncertainty as the model struggles to determine whether a test case from within the overlap is seen or unseen. Further, these generative methods suffer in scenarios with sparse training samples. The models struggle to learn the distribution of high dimensional visual features and, therefore, fail to capture the most discriminative inter-class features. To address these issues, in this paper, we propose a novel framework that leverages dual variational autoencoders with a triplet loss to learn discriminative latent features and applies the entropy-based calibration to minimize the uncertainty in the overlapped area between the seen and unseen classes. To calibrate the uncertainty for seen classes, we calculate the entropy over the softmax probability distribution from a general classifier. With this approach, recognising the seen samples within the seen classes is relatively straightforward, and there is less risk that a seen sample will be misclassified into an unseen class in the overlapped region. Extensive experiments on six benchmark datasets demonstrate that the proposed method outperforms state-of-the-art approaches.

Keywords: Generalized zero shot learning · Image classification · Transfer learning · Triplet network

1 Introduction

Object recognition has seen remarkable advancements since the resurgence of deep convolutional neural networks [9,13]. However, its alluring performance

© Springer Nature Switzerland AG 2021
M. Qiao et al. (Eds.): ADC 2021, LNCS 12610, pp. 139–151, 2021.
https://doi.org/10.1007/978-3-030-69377-0_12

is expensive, with a typical recognition model requiring an inordinate amount of labelled data for training. Without high-quality annotated data, supervised learning breaks down with no way to ensure that a model will be able to predict, classify, or otherwise analyze the phenomenon of interest with any accuracy [20–22,28,30]. Beyond the obvious problems with constructing a robust model, crossing the desert of available data to scale up recognition systems is an arduous, if not impossible, journey.

As a method for overcoming this challenge, zero-shot learning (ZSL) [19,27,32] has been hailed as something of a caravan. With ZSL, models are trained on a small amount of supervised data, i.e., seen classes, where they learn to "mix and match" the features they know to classify unseen objects. This learning paradigm is a subfield of transfer learning, akin to taking knowledge learned from the source domain and using it to complete a task in the target domain [29]. Generalized zero-shot learning (GZSL) [4,6,16,17,25] is a more practical but also more challenging direction of this research that aims to recognise objects in the target domain, i.e., unseen classes while still being able to recognise objects in the source domain, i.e., seen classes. Some state-of-the-art GZSL approaches [5,10,12,14,18,34] use generative models, such as generative adversarial nets (GANs) [8], variational autoencoders (VAEs) [11], or different variants of hybrid GAN/VAEs, to find an alignment between class-level semantic descriptions and visual representations. As such, the zero-shot learning problem becomes one of a traditional supervised classification task. However, there are two main limitations still to be overcome with existing GZSL methods. These are **sparse training samples** [34] and **distribution overlaps**. The sparsity of training samples are common in most image datasets, the distribution of high dimensional visual features are thus hard to learn. As such, the generative models fail to capture the most discriminative inter-class features.

Distribution overlap refers to the potential outcome that a generative model trained only on seen classes becomes overfit to those classes. In this circumstance, the distribution of the synthesised features for the unseen classes can partially, or even fully overlap, with the seen class distribution. The result is uncertainty about whether a test case is seen or unseen and, ultimately, a model with sub-par performance on seen classes. In our experiments, for instance, a simple softmax classifier solely trained on the seen classes was able to achieve more than 90% accuracy on the aPaY dataset. But when unseen samples were included, the seen class accuracy dropped to 51.8%.

To address these issues, we developed a discriminative latent feature generation model with an entropy-based uncertainty calibration technique for GZSL. The generative model contains two variational autoencoders (VAEs) with a triplet loss that learn inter-class discriminative latent features. The encoders in the two VAEs synthesise latent feature embeddings; one is dedicated to the visual space, the other to the semantic space. The latent spaces for the two VAEs are constrained to be shared. Hence, both are able to distil the visual and semantic information. The triplet loss helps to avoid the latent features of each class collapsing into very small clusters by training the generation model to distinguish between similar and dissimilar pairs of examples. This approach minimises the distances between generated latent features from the same class and vice versa for samples in different classes. The remaining two decoders reconstruct the visual and semantic representations from the latent features. An uncertainty calibration is performed during the classification stage, where the latent features of the seen classes are embedded via the visual feature encoder, while the latent features of the unseen classes are generated with the semantic encoder. A general classifier is then constructed over both the seen and unseen classes to classify the latent features. During the testing phase, the general classifier calculates the probability entropy over the seen classes from a softmax output. The probability entropy is a measure of the uncertainty as to whether the sample is seen or unseen. Samples are deemed seen if the entropy value falls below a tuned threshold, and a simple visual classifier is then trained on all the definitively seen samples. The rest of the uncertain test samples are fed into the general classifier to make the final predictions.

In summary, the contributions of this paper are therefore as follows:

- We propose a novel technique for GZSL that exploits an entropy-based uncertainty calibration (EUC) to alleviate issues with distribution overlaps, by training cascade classifiers. Unlike conventional probability calibration techniques that tune a softmax temperature, we propose probability entropy as a measure of confidence that a sample belongs to a seen class.
- A novel generative metric learning (GML) paradigm is presented that leverages dual variational autoencoders with a triplet loss to synthesise latent features. The triplet loss helps to capture the inter-class discrimination in scenarios with sparse training samples.
- Extensive experiments on six benchmark datasets demonstrate the superior performance of the proposed method in both conventional and generalized ZSL against the current state-of-the-art methods.

2   Related Work

Recent state-of-the-art approaches to GZSL with generative models have achieved promising performance. Generative models can synthesise an unlimited number of fake features from side information on the novel classes, e.g., semantic attributes. With these synthesised features, ZSL problems become a relatively straightforward supervised classification task. The two most commonly used generative models are generative adversarial networks (GANs) [8] and variational autoencoders (VAEs) [11]. Often, both models are jointly used to form generative architectures for ZSL tasks. GAZSL [34] leverages Wasserstein GANs (WGAN) [2] to synthesise vivid visual features. Currently, GAZSL models are the state-of-the-art for ZSL tasks with noisy text. CANZSL [5] is a cycle-consistent generative framework for solving domain shift problems and enhancing denoising operations on natural language input. SAE [12] is a semantic autoencoder that learns mappings between the semantic space and the visual space for the encoder to use to translate visual features into the semantic space. The decoder can then

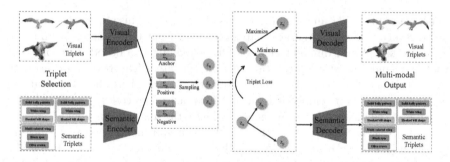

Fig. 1. The proposed GML approach for GZSL. The triplet input consists of an anchor, a positive sample from the target class and a negative sample from one of the different classes. The visual triplet is then fed into the encoder of the visual VAE and, likewise, with the semantic triplets and the encoder of the semantic VAE. The latent vectors are then sampled from the synthesized mean and covariance vectors. Further, a triplet loss is applied to those latent vectors before decoding them into visual and semantic triplets to calculate reconstruction losses.

synthesise unlimited visual features from the semantic information. However, in these methods, the feature generation between the semantic and visual spaces is constrained by asymmetric information. For example, suppose that the wing colour in a bird image is not annotated, so the visual features generated from the semantic labels do not include this information. However, the visual feature extractor can and does extract the wing colour. The two visual representations are now asymmetric, so training a classifier from the generated visual features but testing with extracted visual features may result in sub-optimal performance. With this motivation, our framework incorporates two variational autoencoders with triplet metric regulation to learn latent features that best discriminate between the visual and semantic spaces.

3   Methodology

3.1   Problem Definition

Let T be the set of all possible triplets $\{\tau_a, \tau_p, \tau_n\}$ from the seen classes S, where τ_a and τ_p are from the same category, and τ_n is a negative sample randomly chosen from a different category. Each τ contains a triplet (x, s, y), where x denotes the visual features extracted from an image, s represents the corresponding semantic attributes and y is the associated one-hot class label, respectively. During the evaluation phase, visual features x^u and the corresponding semantic representations s^u are drawn from the unseen classes U. In ZSL, the aim is to predict the class label y^u. By comparison, the aim with GZSL is to predict the class label for the image x from both S and U.

3.2   Dual Generative Model

As shown in Fig. 1, the synthesised features are regularised in the latent space according to the labels by applying a triplet loss. The latent space is formed between the visual and semantic spaces through a dual VAE framework, which includes a visual variational autoencoder v-VAE and a semantic variational autoencoder s-VAE. In terms of the implementation of the dual VAEs, we followed the CADA-VAE approach set out in [26].

The conventional role of VAEs as a generative model leverages the decoder as the generator. By contrast, both the encoders in our dual generative framework work as generators to synthesise embedding vectors in the latent space, while the decoders translate the latent vectors back into visual or semantic features. In other words, the VAE has a cycle structure, providing supervision information to improve the generation ability of the encoders.

The VAE loss for v-VAE and s-VAE can be formulated as:

$$
\begin{aligned}
\mathcal{L}_{v\text{-VAE}} = \ &\mathbb{E}_{Q_v(z_v|x)}[log P_v(x|z_v)] \\
&- \beta_1 D_{KL}[Q_v(z_v|x)||P_v(z_v)],
\end{aligned}
\tag{1}
$$

$$
\begin{aligned}
\mathcal{L}_{s\text{-VAE}} = \ &\mathbb{E}_{Q_s(z_s|s)}[log P_s(s|z_s)] \\
&- \beta_2 D_{KL}[Q_s(z_s|s)||P_s(z_s)],
\end{aligned}
\tag{2}
$$

where Q_v, Q_s denote the encoder networks in v-VAE and s-VAE respectively, and P_v, P_s represents decoder networks. The first terms are the reconstruction loss and the second ones are the KL divergence. β_1 and β_2 are the coefficients of KL divergence for the two VAEs, respectively. z_v and z_s are the latent vectors synthesised from the visual and semantic spaces, respectively. In more detail, the Gaussian distributions $\mathcal{N}_v(\mu_x, \Sigma_x)$, $\mathcal{N}_s(\mu_s, \Sigma_s)$ are the direct output of Q_v and Q_s. z_v and z_s sampled with a reparametrization trick [11] from the two Gaussian distributions.

To make the latent space and intermediate region between the visual and semantic spaces, we use a multi-distribution loss to allow the two Gaussian distributions to approximate each other. In our experiments, we found that a Wasserstein distance gave better performance than other distance functions, such as mean squared error (MSE). Hence, the cross-modal loss is formulated as follows:

$$
\begin{aligned}
\mathbf{W}(\mathcal{N}_v, \mathcal{N}_s)^2 = \ &\|\mu_v - \mu_s\|_2^2 \\
&+ trace(\Sigma_v + \Sigma_s - 2(\Sigma_s^{1/2}\Sigma_v\Sigma_s^{1/2})^{1/2}).
\end{aligned}
\tag{3}
$$

Further, the latent vectors z_v synthesised from the visual space can be decoded into semantic embeddings $\hat{s} \leftarrow P_s(z_v)$ and vice versa for the latent vectors z_s encoded from the semantic space, $\hat{v} \leftarrow P_v(z_s)$. The objective function for optimizing the multi-modal reconstruction is given below:

$$
\mathcal{L}_{mul-recon} = \|\hat{v} - v\|_1 + \|\hat{s} - s\|_1,
\tag{4}
$$

where we use an L1 norm because is it insensitive to outliers. With an L2 norm, too many outliers in the dataset will degrade performance.

3.3   Triplet Regularization

To regularize the latent features for optimal discriminativeness, we have incorporated a triplet loss into the method in addition to the above constraints. As shown in Fig. 1, a triplet loss is enforced in the latent space to ensure a small distance between all objects of the same category and a large pairwise distance between objects different categories. Formally, the triplet loss can be written as follows:

$$\mathcal{L}_{v\text{-}trip} = max(\|Q_v(x_a) - Q_v(x_p)\|^2 - \|Q_v(x_a) - Q_v(x_n)\|^2 + \alpha, 0), \quad (5)$$

where x_a, x_p and x_n are the anchor images, the positive samples and the negative samples, respectively. α represents the margin between the positive and negative pairs, which varies from dataset to dataset. Similar triplet losses $\mathcal{L}_{s\text{-}triplet}$ with the same α are also applied to the latent features z_s generated from the semantic embeddings.

To further improve multi-modal reconstruction and regularize the margin between classes, we propose the following multi-modal triplet objective function:

$$\mathcal{L}_{mul\text{-}trip} = \sum_i^M \sum_j^M \sum_m^M !(i=j=m) max(\mathcal{D}_{i,j,m}, 0\ ), \quad (6)$$

$$\mathcal{D}_{i,j,m} = \|Q_i(i_a) - Q_j(j_p)\|^2 - \|Q_i(i_a) - Q_m(m_n)\|^2 + \alpha, \quad (7)$$

where i, j and m represent the modality M, which can be either visual or semantic. Note that i, j and m cannot be the same modality. This objective function specifies six more triplet loss terms in the latent space.

The superiority of the soft margin in triplet loss over fixed visual centres (aka. visual pivots) as proposed in GAZSL [34] is worth discussing. A visual centre is simply calculated by the mean value of all image features in the same class, and the synthesised visual features from the same class are pushed towards the corresponding visual centre. However, while optimizing the distance between the visual data points, the model may converge into a situation where the samples from different categories overlap. The triplet loss in our framework not only pulls the samples of the same category towards each other, it also pushes negative category objects backwards.

The overall objective function for the proposed approach GML is as follows:

$$\mathcal{L}_{GML} = \mathcal{L}_{v\text{-}VAE} + \mathcal{L}_{s\text{-}VAE} + \lambda \mathbf{W}(\mathcal{N}_v, \mathcal{N}_s)^2 \quad (8)$$
$$+ \mathcal{L}_{mul-recon} + \mathcal{L}_{v\text{-}triplet} + \mathcal{L}_{mul\text{-}trip},$$

where λ denotes the weight of Wasserstein distance.

3.4   Predicting with Uncertainty Calibration

To mitigate catastrophic distribution overlap issues, we further develop a novel entropy-based uncertainty calibration technique to predict the labels of test samples. Once the dual VAE model has been trained, arbitrary instances can be generated in the latent space based on semantic embeddings of the classes. In GZSL,

Table 1. GAZSL accuracy (%) on six datasets. U, S and H represent unseen, seen and harmonic mean, respectively. The best results are formatted in bold.

Methods	aPaY			AWA1			AWA2			CUB			SUN			FLO		
	U	S	H	U	S	H	U	S	H	U	S	H	U	S	H	U	S	H
ALE	4.6	73.7	8.7	16.8	76.1	27.5	14.0	81.8	23.9	23.7	62.8	34.4	21.8	33.1	26.3	13.3	61.6	21.9
SYNC	7.4	66.3	13.3	8.9	87.3	16.2	10.0	90.5	18.0	11.5	**70.9**	19.8	7.9	43.3	13.4	-	-	-
SAE	0.4	80.9	0.9	1.8	77.1	3.5	1.1	82.2	2.2	7.8	54.0	13.6	8.8	18.0	11.8	-	-	-
DEM	11.1	75.1	19.4	32.8	84.7	47.3	30.5	86.4	45.1	19.6	57.9	29.2	20.5	34.3	25.6	-	-	-
GAZSL	14.2	78.6	24.0	29.6	84.2	43.8	35.4	86.9	50.3	31.7	61.3	41.8	22.1	39.3	28.3	28.1	77.4	41.2
GDAN	30.4	75.0	43.4	-	-	-	32.1	67.5	43.5	39.3	66.7	49.5	38.1	**89.9**	**53.4**	-	-	-
CADA-VAE	31.7	55.1	40.3	57.3	72.8	64.1	**55.8**	75.0	63.9	**51.6**	53.5	52.4	43.1	35.4	38.9	51.6	75.6	61.3
Ours	**35.0**	62.7	**44.9**	**60.4**	70.4	**65.1**	55.2	78.9	**64.9**	50.8	55.1	**52.9**	**44.1**	36.8	40.1	**54.0**	**79.0**	64.1

a softmax classifier is trained over both the seen and unseen classes with 200 latent representations for each seen class. Note that the general classifier training data of seen classes are synthesised by the visual encoder Q_v, whereas the latent features for the novel classes are all synthesised from semantic embeddings.

There are three networks involved in predicting test samples: a general classifier f, a seen classifier g and a visual encoder Q_V. Specifically, the general classifier is a softmax classifier that recognizes both seen and unseen samples. Latent features from seen classes and unseen classes need to be provided to train this general classifier. The seen latent features are mapped from the visual samples in the training set by the visual encoder Q_v. Unlimited novel latent features for the unseen classes can be synthesised with the trained semantic encoder Q_s given class embeddings. To predict the labels in conventional GZSL, latent representations are first synthesised by the visual encoder given test the visual feature of the sample. And, once the general classifier is trained, recognizing which class a sample belongs is a straightforward task. When classifying objects into seen classes, almost all unseen objects will have high entropy, whereas approximately half of the seen samples will have lower entropy than the unseen ones. Therefore, entropy is calculated with unnormalized log probabilities from the softmax output. The insight here is that our latent feature generator is trained on the seen classes so whether an object is seen or unseen is more certain as determined by the probability entropy.

Nevertheless, this approach does not prevent overlaps between the seen domain and the unseen domain in the same latent space and, in turn, performance degradation. Therefore, we incorporated an entropy threshold to judge whether an object is likely to be seen; thus, reducing the risk of the general classifier classifying an object into the wrong unseen class. After excluding the uncertain objects, a simple softmax classifier is trained in the seen domain to recognize the low entropy objects. Note that, instead of latent representations, this classifier is trained with visual features, which result in greater than 90% accuracy with seen objects in the testing stage. However, a side effect of the proposed uncertainty calibrator is that performance with unseen classes can degrade slightly since a small portion of novel class samples are improperly classified into seen classes. Nevertheless, this tiny sacrifice of accuracy in the unseen domain is more than offset by the substantial improvement in accuracy in the seen domain (Fig. 2).

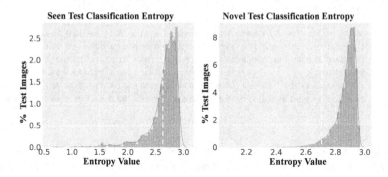

Fig. 2. The probability entropy for the seen and novel test classifications on the aPaY dataset. The white line indicates the entropy threshold (of 2.7) that separates certain seen samples (left) from uncertain ones (right). It is clear that only a few novel test samples were misclassified as seen classes, whereas a large portion of the seen samples were confidently categorized into seen classes.

4   Experiments

4.1   Datasets and Compared Methods

To demonstrate the robustness of our method, we conduct experiments on six benchmark ZSL datasets, including three coarse-grained datasets (aPaY [7], AWA1 [15], AWA2 [31]) and three fine-grained medium-sized datasets (CUB [1], SUN [24] and FLO [23]). We evaluated performance against representative methods proposed over the last few years as well as recent state-of-the-art frameworks. These include ALE [1], SYNC [3], SAE [12], DEM [33], GAZSL [34], GDAN [10], CADA-VAE [26].

4.2   Evaluation Protocol

The metric used to evaluate both the generalized and conventional ZSL tasks was the widely-used average per-class top-1 accuracy. We use harmonic mean as the evaluation criteria to calculate the joint accuracy of the source and target domains because our goal is to ensure good performance in both domains, and an arithmetic mean could be high simply due to stellar performance in one domain. The formula used to calculate the harmonic mean $\mathcal{H}$ is provided below:

$$\mathcal{H} = 2 * (acc_{\mathcal{Y}^{tr}} * acc_{\mathcal{Y}^{ts}})/(acc_{\mathcal{Y}^{tr}} + acc_{\mathcal{Y}^{ts}}). \tag{9}$$

where $\mathcal{Y}^{ts}$ and $\mathcal{Y}^{tr}$ denote the accuracy of seen and unseen classes, respectively.

4.3 GZSL Results

The results of the GZSL task for all methods are provided in Table 1, showing that our method outperforms the others on most of the datasets. An analysis of these results confirms that the uncertainty calibrator improved model performance with seen objects to a statistically significant degree. At the same time, the results for the unseen classes equal the state-of-the-art. For example, on the aPaY dataset, the visual classifier for the source domain yielded 98% accuracy on the training set of seen objects, and 91% accuracy on the test set of seen objects. For the samples certainly from seen classes, the visual classifier is applied to classify them into seen classes. In conventional GZSL settings, seen class accuracy is merely 51.8% with a general classifier but, with the uncertainty calibrator, it surges dramatically to 62.7%. The minor trade-off in performance with the novel classes (0.2%) is entirely worthwhile for an improvement of 41.0% to 44.9% in terms of the harmonic mean. Figure 3 shows the statistics of the classification entropy for the test samples from the aPaY dataset. Notably, the entropy of samples in the seen and novel classes have different distributions. So, when we set the entropy threshold to 2.7, a large amount of the seen test samples fall below the threshold, which means they were confidently classified into the source domain. Conversely, almost all of the novel test samples fall above the threshold are were easily classified into unseen classes. Even with the few misclassifications that slipped through, conventional GZSL approaches struggle in these circumstances as the results show.

Table 2. Conventional ZSL accuracy (%). The best results are formatted in bold.

	CUB	aPaY	AWA1	AWA2	SUN	FLO
ALE	54.9	39.7	59.9	62.5	58.1	48.5
SYNC	55.6	23.8	54.0	46.6	56.3	-
SAE	33.3	8.3	53.0	54.1	40.3	-
GDAN	39.3	30.4	-	32.1	38.1	-
DEM	51.7	35.0	**68.4**	67.1	61.9	
GAZSL	55.8	**41.1**	68.2	**70.2**	61.3	60.5
Ours	**61.73**	39.1	65.7	66.0	**63.5**	**67.2**

4.4 Conventional ZSL Results

As a further analysis, we are curious to see how our framework compared with conventional ZSL, where only objects in unseen classes need to be classified. We use our dual VAE model to synthesise a fixed amount of latent representations (400 was optimal in this case), given novel semantic embeddings in n classes, then train an n-way classifier on the supervised data. Again, average per-class top-1 accuracy is the evaluation metric. The results, shown in Table 2, confirm that our approach not only performs well with GZSL tasks but also delivers state-of-the-art performance in conventional ZSL. It is also worth mentioning that the proposed model outperformed all state-of-the-art methods on the CUB, SUN and FLO datasets. The result with the SUN dataset is particularly promising since this dataset has more than 700 categories with only 20 images per category, and accuracy still reached 63.5%.

4.5 Ablation Study

Our ablation study covers the effects of the basic dual VAE framework, the triplet loss and the entropy-based uncertainty calibrator. We train a variant of our model without the entropy-based uncertainty calibrator (EUC), and another without either the uncertainty calibrator or triplet loss and compared performance with the complete framework. The results appear in Table 3. It is clear that each component positively and significantly contributes to performance. The triplet loss improves accuracy by around 2%, while the uncertainty calibrator improves performance on the seen classes significantly as discussed above.

Table 3. Ablation study. Effects of different components on GZSL performance (%) on datasets aPaY and FLO.

Model		w/o EUC & Trip	w/o EUC	Complete
aPaY	U	31.7	**35.1**	35.0
	S	55.1	49.2	**62.7**
	H	40.3	41.0	**44.9**
FLO	U	51.6	**54.2**	54.0
	S	75.6	77.5	**79.0**
	H	61.3	63.8	**64.1**

4.6 Latent Space Distribution Analysis

To further verify the triplet loss regularisation, we produce a t-SNE visualization with the AWA dataset. Figure 3 visualizes the unseen visual samples from the test set and the corresponding latent features generated by the visual encoder. The distributions of the visual features and the encoded latent features are approximate to each other, which proves that the visual encoder preserves the most visual distribution information for the unseen visual samples. Apart from avoiding the information loss problem, the encoded visual features are more discriminative. For example, the visual features of Bat (marked in green) are mixed with Rabbit (marked in red) and Persian cat (marked in gray). This problem is alleviated by the visual feature encoder. It is particularly noticeable in Fig. 3(a)

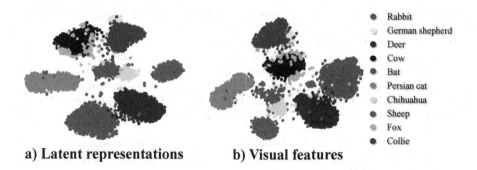

a) **Latent representations**   b) **Visual features**

Fig. 3. t-SNE visualisation of the distribution of encoded latent features from visual features (left) and visual features (right). It can be seen the encoded visual features are even more discriminative than extracted the visual features. (Color figure online)

that the data points from Bat are more centralized than the original visual features. Moreover, since the negative class samples are forced to separate during training with triplet loss, the distribution is more inter-class discriminative.

5   Conclusion

In this paper, we proposed a novel framework for GZSL that uses dual VAEs with a triplet framework to learn discriminative latent features. An entropy-based uncertainty calibration minimizes overlapping areas between seen and unseen classes that can lead to performance degradation in seen classes by leveraging the entropy of the softmax probability over those seen classes. Consequently, the classes deemed to be seen with high confidence can be used to train a classifier with a supervised model over seen classes. This feature generation framework proved to be effective with both GZSL and conventional ZSL tasks. However, despite the remarkable results achieved by the framework, the important threshold for the uncertainty calibration must be tuned for each individual dataset. Developing a mechanism for automatically tuning this hyper-parameter is a potential direction for future work.

Acknowledgment. This work was partially supported by ARC DP190101985, ARC DE200101610, Sichuan Science and Technology Program under Grant (2020YFG0080) and the National Natural Science Foundation of China (No. 62002085).

References

1. Akata, Z., Perronnin, F., Harchaoui, Z., Schmid, C.: Label-embedding for image classification. IEEE TPAMI **38**(7), 1425–1438 (2015)
2. Arjovsky, M., Chintala, S., Bottou, L.: Wasserstein generative adversarial networks. In: ICML, pp. 214–223 (2017)

3. Changpinyo, S., Chao, W.L., Gong, B., Sha, F.: Synthesized classifiers for zero-shot learning. In: CVPR, pp. 5327–5336 (2016)
4. Chao, W.-L., Changpinyo, S., Gong, B., Sha, F.: An empirical study and analysis of generalized zero-shot learning for object recognition in the wild. In: Leibe, B., Matas, J., Sebe, N., Welling, M. (eds.) ECCV 2016. LNCS, vol. 9906, pp. 52–68. Springer, Cham (2016). https://doi.org/10.1007/978-3-319-46475-6_4
5. Chen, Z., Li, J., Luo, Y., Huang, Z., Yang, Y.: CANZSL: cycle-consistent adversarial networks for zero-shot learning from natural language. In: WACV, pp. 874–883 (2020)
6. Chen, Z., Wang, S., Li, J., Huang, Z.: Rethinking generative zero-shot learning: an ensemble learning perspective for recognising visual patches. In: ACM MM, pp. 3413–3421 (2020)
7. Farhadi, A., Endres, I., Hoiem, D., Forsyth, D.: Describing objects by their attributes. In: CVPR, pp. 1778–1785. IEEE (2009)
8. Goodfellow, I., et al.: Generative adversarial nets. In: NeurIPS (2014)
9. He, K., Zhang, X., Ren, S., Sun, J.: Deep residual learning for image recognition. In: CVPR, pp. 770–778 (2016)
10. Huang, H., Wang, C., Yu, P.S., Wang, C.D.: Generative dual adversarial network for generalized zero-shot learning. In: CVPR, pp. 801–810 (2019)
11. Kingma, D.P., Wellin, M.: Auto-encoding variational Bayes. In: ICLR (2014)
12. Kodirov, E., Xiang, T., Gong, S.: Semantic autoencoder for zero-shot learning. In: CVPR, pp. 3174–3183 (2017)
13. Krizhevsky, A., Sutskever, I., Hinton, G.E.: ImageNet classification with deep convolutional neural networks. In: NeurIPS, pp. 1097–1105 (2012)
14. Kumar Verma, V., Arora, G., Mishra, A., Rai, P.: Generalized zero-shot learning via synthesized examples. In: CVPR, pp. 4281–4289 (2018)
15. Lampert, C.H., Nickisch, H., Harmeling, S.: Attribute-based classification for zero-shot visual object categorization. IEEE TPAMI 36(3), 453–465 (2013)
16. Li, J., Jing, M., Lu, K., Zhu, L., Yang, Y., Huang, Z.: Alleviating feature confusion for generative zero-shot learning. In: ACM MM, pp. 1587–1595. ACM (2019)
17. Li, J., Jing, M., Lu, K., Zhu, L., Yang, Y., Huang, Z.: From zero-shot learning to cold-start recommendation. In: AAAI, vol. 25, p. 30 (2019)
18. Li, J., Jing, M., Zhu, L., Ding, Z., Lu, K., Yang, Y.: Learning modality-invariant latent representations for generalized zero-shot learning. In: ACM MM, pp. 1348–1356 (2020)
19. Luo, C., Li, Z., Huang, K., Feng, J., Wang, M.: Zero-shot learning via attribute regression and class prototype rectification. IEEE TIP 27(2), 637–648 (2017)
20. Luo, Y., Huang, Z., Wang, Z., Zhang, Z., Baktashmotlagh, M.: Adversarial bipartite graph learning for video domain adaptation. In: ACM MM, pp. 19–27 (2020)
21. Luo, Y., Huang, Z., Zhang, Z., Wang, Z., Baktashmotlagh, M., Yang, Y.: Learning from the past: continual meta-learning via Bayesian graph modeling. arXiv preprint arXiv:1911.04695 (2019)
22. Luo, Y., Wang, Z., Huang, Z., Baktashmotlagh, M.: Progressive graph learning for open-set domain adaptation. In: ICML, pp. 6468–6478. PMLR (2020)
23. Nilsback, M.E., Zisserman, A.: Automated flower classification over a large number of classes. In: 2008 Sixth Indian Conference on Computer Vision, Graphics & Image Processing, pp. 722–729. IEEE (2008)
24. Patterson, G., Hays, J.: Sun attribute database: discovering, annotating, and recognizing scene attributes. In: CVPR, pp. 2751–2758. IEEE (2012)
25. Rahman, S., Khan, S., Porikli, F.: A unified approach for conventional zero-shot, generalized zero-shot, and few-shot learning. IEEE TIP 27(11), 5652–5667 (2018)

26. Schonfeld, E., Ebrahimi, S., Sinha, S., Darrell, T., Akata, Z.: Generalized zero-and few-shot learning via aligned variational autoencoders. In: CVPR (2019)
27. Shen, F., Zhou, X., Yu, J., Yang, Y., Liu, L., Shen, H.T.: Scalable zero-shot learning via binary visual-semantic embeddings. IEEE TIP **28**(7), 3662–3674 (2019)
28. Simonyan, K., Zisserman, A.: Very deep convolutional networks for large-scale image recognition. arXiv preprint arXiv:1409.1556 (2014)
29. Wang, W., Zheng, V.W., Yu, H., Miao, C.: A survey of zero-shot learning: settings, methods, and applications. ACM Trans. Intell. Syst. Technol. (TIST) **10**(2), 13 (2019)
30. Wang, Z., Luo, Y., Huang, Z., Baktashmotlagh, M.: Prototype-matching graph network for heterogeneous domain adaptation. In: ACM MM, pp. 2104–2112 (2020)
31. Xian, Y., Lampert, C.H., Schiele, B., Akata, Z.: Zero-shot learning-a comprehensive evaluation of the good, the bad and the ugly. IEEE TPAMI **41**, 2251–2265 (2018)
32. Yang, Y., Luo, Y., Chen, W., Shen, F., Shao, J., Shen, H.T.: Zero-shot hashing via transferring supervised knowledge. In: ACM MM, pp. 1286–1295. ACM (2016)
33. Zhang, L., Xiang, T., Gong, S.: Learning a deep embedding model for zero-shot learning. In: CVPR, pp. 2021–2030 (2017)
34. Zhu, Y., Elhoseiny, M., Liu, B., Peng, X., Elgammal, A.: A generative adversarial approach for zero-shot learning from noisy texts. In: CVPR, pp. 1004–1013 (2018)

A Real Time Analysis of Offensive Texts to Prevent Cyberbullying

Sonia Khetarpaul(✉)📵, Dolly Sharma📵, Mayuri Gupta📵,
and Vaibhav Gautam

Department of Computer Science and Engineering, Shiv Nadar University,
Gautam Buddha Nagar, India
{sonia.khetarpaul,dolly.sharma,mg153,vg459}@snu.edu.in

Abstract. Cyberbullying is one of the leading causes of mental health issues in the younger population, often leading to depression, stress, and suicidal tendencies. Often it is observed that timely intervention can bring awareness towards unintentional cyberbullying. This paper presents an approach to the prevention of cyberbullying via social networking. The objective of this work is real-time detection of the degree of offensiveness and alerting the user typing the message. We also propose corrective measures, by displaying an alternative word for any offensive word that is typed in the suggestion box in real-time. This enables the user to prevent unintentional cyberbullying. The proposed solution is integrated into an app, that displays the relevant statistics as well as visualization of the user typing pattern. The model to detect the offensiveness percentage can achieve 97.77% accuracy and is the backbone of the entire approach.

1 Introduction

Social media has become an integral part of the lives of many of us. It is used by people to share their thoughts, images, videos, and be in constant touch with anyone in the real world. Roughly, 3.5 Billion people around the world are using social media in one way or the other [1]. Over time some Social Media Platforms like Facebook, YouTube, etc. have developed sensitivity towards reported objectionable content and abusive posts, however, there is no real-time tracking of the personal text messages a person can send to another person. With the increasing number of such social networking platforms influencing a wider population, instances of cyberbullying have become more common; and is deeply affecting the younger generation mentally [2]. Cyberbullying is a major cause of depression, insecurity, stress, anxiety, insomnia, and other mental disorders among users. 59% of teens report that they have been bullied online and that has affected their daily lives. Around 20 % of the students have reported reluctance in going to school. 5% of the cyberbullied users have reported self-harm and 3% of them report attempted suicide [3].

Offensive texts are very subjective and often difficult to identify. Most of the cases of cyberbullying are unintentional but the receiver gets offended through

© Springer Nature Switzerland AG 2021
M. Qiao et al. (Eds.): ADC 2021, LNCS 12610, pp. 152–165, 2021.
https://doi.org/10.1007/978-3-030-69377-0_13

them. We consider the use of hateful or abusive language, sexually explicit words, racist, violence, and other terms that might be degrading in some sense as offensive. Unless explicitly reported these cases will remain unidentified. Real-time detection of offensive texts to prevent instances of cyberbullying is thus a topic of urgent interest today.

1.1   State-of-the-Art in Detecting Offensive Text

a) **Keyword Censoring:** In this method, all offensive words appearing in a text message are censored. Offensive words can be removed completely from the sentence or partially replaced with '*'(e.g.- c****) or completely replaced with some family-friendly word. Even though the method is simple it doesn't give the desired results [4]. If the words are removed completely from the sentence, the sentence completely loses its meaning. If the words are partially replaced with '*' it is very easy to guess the word. The idea of replacing the offensive word with some family-friendly words seems very useful but it is very difficult to accurately do that and can cause even more problems.

b) **Content Control:** Content-control processes are commonly deployed at the user's side or ISP side to prevent each user from seeing inappropriate content material on the Web. The filtering in this technique is based on certain criteria like URL address, the occurrence of offensive words, and as well as topic classification [5]. This technique is considered to be too coarse-grained to be used in many online platforms. This is because a sentence may contain offensive as well as non-offensive words. So, this technique will remove all the non-offensive words falsely which might be important for the current user.

c) **Manual Filtering:** A manual filtering approach is believed to provide the best filtering result. In this method, the messages of the user are reviewed by the administrator of the community before posting anything on the website [6]. Manually filtering takes a long time and it is even a tedious job to do. The results of this method depend on how the administrator views subjective content. Even in this method, one can expect a delay between the submission of the text and the posting of the text.

1.2   Our Novel Approach and Contributions

In the proposed method we use the best of all the above approaches without inheriting any of their shortcomings. In our proposed method we do not remove words from the user's message, instead, we ask the user themselves to review their messages based on the degree of the offensiveness before sending it to another person. The sender himself does the manual filtering and replaces the offensive words with one of the suggested words which best suits the sentence semantically.

We create a novel application shown in Fig. 1, which notifies the user of the degree of offensiveness in the text as it is being typed by them and suggest alternate words. The proposed model can be integrated with any social media platform. The key contributions of the proposed work are:

- Real-time detection of the degree of offensiveness in the text being typed.
- The corrective measures, by displaying an alternate word in the suggestion box for any offensive word that is typed in real-time.
- The display of the relevant statistics as well as the visualization of the user typing pattern.

Fig. 1. Screenshot of proposed mobile application

The rest of the paper is organized as follows. Section 2 presents the existing work related to the detecting cyberbullying and filtration of offensive content in social media platforms. Section 3 presents an overview of the proposed approach. Section 4 focuses on the aspects of implementation and development of the application. Section 5 focuses on the detection of Real-time offensive text. The results are discussed in Sect. 6. Finally, we conclude the paper in Sect. 7.

2   Related Work

Cyberbullying and cyber abusing detection have been important topics over the last few years. The various researchers have mainly focused on textual data analysis [12,13]. There has been an attempt to use the Fuzzy logic to filter out the input text for preparing for classification and then detection of cyber abuse using the various genetic algorithms [11]. Some of the researchers have worked with machine learning techniques on collected textual data with an accuracy of 78.5% using the decision tree learner and instance-based learner [14]. One of the methods adopted is to apply common sense reasoning on the textual data which uses common sense reasoning for cyber abuse detection [26]. Some methods use Lexical Syntactic Feature which detects cyber abusiveness using the semantic analysis of the sentences [15].

Some researchers have examined facial expressions to detect cyber abuse [17–19]. *Balett et al.* had achieved 85% accuracy for the detection of the pain using computer vision by measuring the facial movements and pattern recognitio [17]. *Bonanno et al.*, in a psychological study, examine disclosure-nondisclosure of early life sexual abuse with nonverbal expressions of emotion in faces of subjects [20]. Besides, *Christani et al.*, appoint Social Signal Processing based on video surveillance for identifying non-verbal cues which includes facial expressions, gazing, and frame postures to relate them with context dependent activities [21]. Authors in [27] detected complicated facial expressions that indicate the purpose of abuse. In the realm of connecting abusive conduct with human emotions, studies show that there are strong correlations between anger and giving a reason to abuse [22–25]. *Wang et al.*, experimented with 464 young Chinese adults and applied the social-cognitive model alongside with fashionable aggression model [24]. *Bosworth et al.*, in [23] look at that anger becomes a powerful predictor of abusive behavior. They conclude that excessive ranges of anger are associated notably with the very best ranges of abuse. *Hussain et al* in [28], display that victims of abuse exhibit anger-in phenomenon as opposed to anger-out by perpetrators of abuse. The current natural language approaches don't work for social networks to generalize cyber abusive detection for young people who uses different short forms [16]. Current approaches do not have real-time text detection which gives high accuracy. But our approach can also deal with most of the short forms and displays a real-time offensive percentage. There are many challenges in using the already collected data and then detecting cyber abuse. So, we are using the possibility of real-time detection of abusiveness in the sentence. However, there are many shortcomings of this approach which are already described above. In our proposed method we do not remove words from the user's message, instead, we ask the user themselves to review their messages based on the percentage of the offensiveness before sending it to another person. Our proposed approach is the mix of all the above approaches without inheriting any of their shortcomings as the sender himself does the manual filtering and replaces the offensive words with one of the suggested words which best suits the sentence semantically.

3   Proposed Architecture

The architecture of our proposed offensive text detection model is shown in Fig. 2. The model has two parts (differentiated using two different colors in Fig. 2), (i) Development and implementation of the model. (ii) Real-time offensive text detection. Details of each part are described in Sects. 4 and 5, respectively.

As shown in the proposed architecture (Fig. 2), the messages are first extracted from the database and preprocessed. The data prepossessing including stemming, lemmatization, stop word removal, the lower casing is performed on the data. Next, the sentences are tokenized into various tokens while keeping only the most frequently occurring words in the text corpus. To keep every sentence

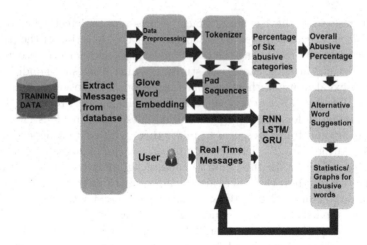

Fig. 2. Architecture of the proposed model

of the same length, the sequences are padded. To convert the words present in the sentence to the numerical values which can be then further be fed into our training model, we have used GloVe word embedding. Now the real-time messages that the user types in along with training data are fed into our RNN model to predict the percentage of six abusive categories and then predict the overall percentage of abusiveness in the sentence. Then, our model provides alternative words that can be used instead of the detected abusive words to decrease the percentage of abusiveness in the sentence. Our application also displays the graphs and statistics of abusive words that have been used in the past by that user. The details about the development and implementation of the proposed architecture are described in the next section.

4   Application Development and Implementation Aspects

As described earlier, the user will be entering their text using the keyboard. The following steps are followed to analyze this text to predict the percentage score of the abusiveness in their sentence and give the alternative suggestions to those abusive words.

4.1   Data Preprocessing

Data/Text preprocessing is an important and first step before feeding any data into the Machine learning model. The text preprocessing steps include the following:

– Lower Casing: In this step, we have reduced all the letters of the text into lower case letters. It will be easy to match the query of uppercase letters to the vocabulary of lower case letters.

- Stop words Removal: We have used stop word removal as a step to remove those words that appear to be of less importance when a query is searched and should be removed from the vocabulary that will help to train our model better.
- Stemming: Stemming often includes the removal of derivational affixes. There are many stemming algorithms but for this, we have used Porter's Stemmer algorithm. It is believed to be based on the idea that all the suffixes in the English language are made up of a combination of smaller and simpler suffixes.
- Lemmatization: In this step, we used lemmatization to reduce similar semantic words to a single word. The process grouped the distinct inflected forms of a word so they can be analyzed as a single item.

4.2 Input Transformation

After the data preprocessing step, the next step is to transform our input to be used by various machine learning models. This includes vectorizing the input to perform Natural Language Processing techniques.

Noise Removal: While extracting the data, we observed that data is noisy, that is, it contains text that is not relevant to the context. During this step, all types of noise entities present in the text are removed. The approach that we used for noise removal is to prepare a dictionary of noise entities. Then we iterate every word by tokens or words and keep on removing those words that are present in the noise dictionary.

Representation: Creation of Sequences: The next step is to convert the input real-time text as well as the dataset into a series of numbers that can be easily fed into the model for training and getting the predictions. The following steps are performed to create the sequences:

- **Tokenizer:** A tokenizer is a utility function that is used to split a particular sentence into words. We have used the Keras Tokenizer function To keep the pre-specified number of words in the text, num_words parameter is used. This is useful as we don't want our models to get a lot of noise by considering the infrequent words. The words that are left by num_words are mostly misspelled words.
- **Pad Sequence:** Generally, the model expects that every sequence(training example) should be of the same length i.e. same number of words or tokens. This can be controlled using the maxlen parameter. The train and test data contains a list of numbers. Each and every list has the same length. The word_index is the dictionary of most frequently occurring words in the text corpus.

4.3   Embedding Enrichment

Word embedding is considered to be one of the most popular representations of the document vocabulary. It can capture the context of a word in a document, semantic and syntactic similarity, relation with other words, etc. Word embedding is the representation of a particular word in the form of vectors.

GloVe: The GloVe is primarily based on matrix factorization strategies on the word-context matrix [7]. The algorithm first constructs a huge matrix of (words x context) co-occurrence information, i.e. For each "phrase" (the rows), GloVe, remember how frequently this phrase is seen in some "context" (the columns) in a big corpus. For our experimentation, the GloVe word embedding method is used. Using the GloVe embedding file, the input sentence is vectorized so as to convert words present in the sentence to numerical values which can be then further fed into our training NLP model. The GloVe is used because it creates a global co-occurrence matrix by estimating the probability a given word will co-occur with other words. This presence of global information makes it ideally works better.

4.4   Recurrent Neural Network

Recurrent means that the output at the current time step is the input to the next time step. At every element of the sequence, the model considers the current input and also considers what it remembers about the preceding elements [8]. Since in real-time detection of offensive words, the network should remember the previous words or the sentence, hidden layers of the RNN do remember such sequences. Since these layers don't learn, RNN's have the capability that can forget longer sequences as they will have a tough time to carry records from earlier time steps to later ones, thus having a short-term memory. **LSTM's** and **GRU's** were discovered to deal with the problem of short-term memory. These gates can study which information in a chain is important to maintain or throw away. By doing that, it can skip relevant records down the lengthy chain of sequences to make predictions.

Long Short-Term Memory: LSTM keep only the important information and throw the rest, passing only the relevant information for further processing to make the predictions [9]. So, we used Long Short-Term Memory(LSTM) because we want to pass only the relevant information that is the abusive words that will play an important role in predicting the score of the sentence abusiveness. Also, since it is based on the short-term memory, so it will only need to predict the score based on current input as a text not the previous texts that a person wrote.

Bidirectional LSTM is being used which is an extension of LSTMs that can improve model performance on sequence classification problems [10]. Then, GlobalMaxPool1D() is used to reduce the dimensionality of the feature maps that has been output by some layer, to replace Flattening and sometimes even Dense layers in our classifier. The first dense layer outputs 50 as the hidden nodes with the first activation function and then the next dense layer outputs 6 output nodes with the second activation function. So, we have tried out the various combinations of LSTM and these are shown in Table 1. The model is then compiled using the loss function as 'binary_crossentropy' and optimizer as 'adam'.

Table 1. Long short-term memory results

Model name	TRAIN ACCURACY (%)	TEST ACCURACY (%)
LSTM (ReLu and sigmoid)	98.26	97.718
LSTM (tanh and sigmoid)	98.35	97.777
LSTM (ReLu and tanh)	97.23	96.781
LSTM Convolution (tanh and sigmoid)	98.07	97.597
LSTM Convolution (ReLu and sigmoid)	98.25	97.724

Gated Recurrent Unit: It is a new generation of Recurrent Neural Networks, works similar to the LSTM. Gated Recurrent Unit(GRU) uses only the hidden state to transfer the information and get rid of the cell states. Unlike LSTM, it consists of the most effective three gates and does not hold an Internal Cell State. The procedure for training GRU is the same as LSTM by using a bidirectional model, GlobalMaxPool1D, and dense layers combination as it was trained with LSTM. Everything remains the same but instead model is GRU rather than LSTM. The results are shown in Table 2.

Table 2. Gated recurrent unit results

Model Name	TRAIN ACCURACY (%)	TEST ACCURACY (%)
GRU (ReLu and sigmoid)	98.02	97.178
GRU (tanh and sigmoid)	98.12	97.279
GRU (ReLu and tanh)	97.021	95.793
GRU Convolution (tanh and sigmoid)	97.98	96.159
GRU Convolution (ReLu and sigmoid)	98.06	96.257

After training the model and testing on test data, the best model achieved is the model LSTM with tanh and sigmoid as the activation function with the accuracy rate of 97.777%. After training and testing the model, we predicted the values for our sentence that the user is entering in real-time. The model will return 6 values i.e. the score of threat, severe toxic, obscene, toxic, hate, and insult between 0 and 1. These values will be returned to the user's mobile phone through the API that we built using the flask. The values will be returned in the form of JSON. So we will parse the JSON and then the one with the maximum will be the predicted score of the abusiveness of his/her sentence.

4.5   Alternative Word Suggestion

Abusive words can be replaced by alternative words thus reducing the degree of abusiveness in a sentence. To suggest to the user some alternative words that can be used instead of the abusive words, we used spaCy. It is an open-source software library for natural language processing.

After the real-time analysis and computation of the percentage of the abusiveness in the sentence, the abusive words are identified. The dataset of abusive words is downloaded from Kaggle and then in a sentence, each word is checked across all the words in the dataset. If it is found, then that word is marked abusive that the user can use and alternate options are provided without altering the semantic of the sentence the same and reducing the percentage of abusiveness as well. All the synonyms of that abusive word are found using the nltk corpus WordNet.

Now, for each synonym, two things need to be calculated. First, the similarity is calculated between that synonym and the abusive word. Second, the percentage of abusiveness for that synonym. The best synonym is chosen such that the percentage of abusiveness is low as compared to others and also should have a higher similarity between the abusive word and that synonym. This best synonym is then suggested to the user to be used in the sentence. The percentage of abusiveness is calculated using our best model trained. The similarity between a synonym and the abusive word is calculated using the spaCy for vector similarity. Each word is said to have a vector representation that is being learned by contextual embeddings (Word2Vec), which are trained on the corpora. The similarity is being determined by comparing the two-word vectors or "word embeddings", multi-dimensional meaning representations of a word. Word vectors are said to be generated using an algorithm like word2vec. The similarity used by spaCy is the cosine similarity.

5   Real-Time Offensive Text Detection

We created an android keyboard application, that makes use of the model described in Sect. 4. The user has to first install and activate the keyboard from the settings. This is checked every time the user installs the app and if the keyboard is not activated then the app prompts the user to activate it. An API is hosted on the server which answers queries based on the best model chosen after experimentation. The training of the model and the hosting of the final model in the form of API is done on the server because it is much faster on the server compared to the mobile phone.

Whenever a user types anywhere in his/her phone, the text that is being typed is sent to the server for processing. Every time the user presses space in the phone keyboard the query is sent to the server. The model processes the query and returns the output in the following 6 categories in a JSON format: i) Toxic, ii) Severe_toxic, iii) Obscene, iv) Threat, v) Insult, and vi) Hate. In the user's mobile phone, the JSON is processed to extract 6 different percentage values, one for each category. The maximum of these percentages is chosen as the final offense percentage. This percentage is displayed in the top left corner just above the keyboard as shown in Fig. 1. The percentage keeps updating in real-time as the user keeps typing. Whenever the degree of offensiveness reaches a particular threshold, we aim to highlight the offensiveness word and suggest the alternative word that can be used instead of the original word.

A list of all the words/sentences that are considered as offensiveness has been added to the app along with their alternative. The check happens in real-time. As soon as the user taps a space the word that was typed last is checked for offensiveness. If the word is considered offensiveness, an alternative of the word is displayed just to the right side of the offensiveness percentage. The user can choose to use the alternative word or keep using the original word. The focus is on providing good alternatives rather than forcing the user to not use a particular word, which might irritate her/him. After the user makes the decision the degree of offensiveness is calculated again in real-time based on the new sentence. We also keep track of the user's usage over a while. We record each of the words typed by the user and create a table of the 5 most frequently typed words by the user. The aim is to make the user aware of his voluntarily/involuntarily habits of using some words that might be offensiveness to use. These results can be seen on the statistics page in the app.

The app also contains the visualization of the data on usage of the offensiveness words by the user with respect to time which is being stored while they use those offensiveness words. As shown in the Fig. 3, a line graph is displayed based on the percentage of offense in user's texts over a period of time. On the X-axis we have time, whereas on the Y-axis we have the offense percentage. This is done so that the user can easily see the improvement in his text patterns over time.

6   Performance Evaluation and Results

We evaluate performance of our proposed approach through real experimentation. First, we explain the dataset used for experiments followed by discussion of efficiency of real-time offensiveness text detection module.

6.1   Dataset Used

The dataset is collected from Kaggle [29] which comprises total 312735 text messages. Out of the total, 159571 text messages are selected as the training set and 153164 text messages as the test set for validation. The training dataset contains the text message and the six different categories: toxic, severe_toxic, obscene, threat, insult and indentity_hate. The number of messages in each category is defined in Table 3.

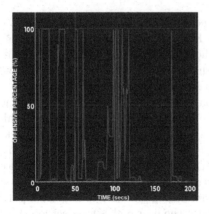

Fig. 3. Screenshot of visualization section

Table 3. Dataset statistics

Category	# messages
toxic	15294
severe_toxic	1595
obscene	8449
threat	478
insult	7877
indentity_hate	1405

6.2   Experimental Setup and Results

After analyzing various models (as described in Tables 1 and 2), we selected LSTM with tanh and sigmoid activation function with the help of exhaustive experiments we could attain an accuracy of 97.77%. This model is chosen to be hosted on the API. As mentioned before, the model returns the output in the six categories. Here are some of the examples, shown in Table 4, of the results obtained from the user.

The examples are chosen by observing the common usage of offensiveness words. It was noted that the word 'fuck' and 'bitch' are some of the most commonly used offensiveness words. The model correctly identifies good and bad sentences and assigns a percentage in all the categories. Percentages are pretty similar to what someone would manually annotate. This is expected given the high accuracy of the model being used.

Table 4. Offensiveness percentage of different sentences as predicted by the model

Text	TOXIC%	Severe_toxic%	Obscene%	Threat%	Insult%	Hate%
You are a bitch	98.6	42.6	97.6	4.5	89.11	8.1
Oh fuck	97	38.8	98.1	3.5	70.8	2.7
Bullshit	94.8	18.8	94.3	2.19	62.4	2.2
Fuck you bitch	99.2	70	98.9	7.3	91.5	11.1
You are awesome	22.2	0.2	2.2	0.5	5.3	0.2

Now the next step after identifying the offense percentage is to suggest the alternative words to the user. Considering the first 3 of the above samples as offensiveness, our model suggests replacing the word 'bitch', 'fuck', and 'bullshit' with 'bad person', 'snap', and 'bullspit' respectively. The alternative words suggested don't have the same meaning as the original words, but the overall meaning of the sentence remains almost the same. Now the offensiveness words are replaced and the updated sentence is again sent to the API to process, results shown in Table 5.

Table 5. Comparison between user's original text and text after applying our suggestions

Original text	Offensiveness(%)	Suggested text	Offensiveness(%)
You are a bitch	98.6	You are a bad person	27.5
Oh fuck	98.1	Oh snap	1.7
Bullshit	94.8	Bullspit	1.4
Fuck you bitch	99.2	Snap you bad woman	7.2

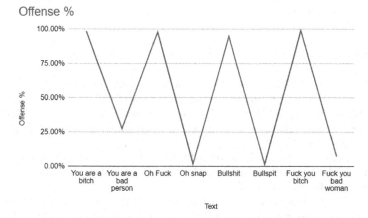

Fig. 4. Graph based on the examples mentioned above

As shown in Table 5, evidently, the offense percentage is reduced without changing the meaning of the sentence. It is also observed that using a particular word, which in normal circumstances doesn't make sense but is able to convey the meaning effectively, reduce the percentage a lot. For example, using snap instead of 'fuck', the feel of surprise, is conveyed almost 100% and the user receiving the message would never know if the person sending the message actually wrote 'snap' or 'fuck'. In some examples, like the first one, where the word 'bitch' is replaced by 'bad woman' the percentage does not reduce to single digits like in other examples. It is because calling somebody a bad person is an offensiveness to some extent but certainly less offensiveness than calling someone a 'bitch'. The entire process is a tradeoff between the choice of words and offense percentage as shown in Fig. 4. The words that change the meaning of the sentence aren't considered as good alternatives but at the same time, the words have to be chosen in such a way that they reduce the offense percentage.

7   Conclusion

The aim of this work is to prevent cyberbullying from using real-time text analysis. We have to build a model that shows satisfactory and promising results. This kind of model can be easily integrated into any keyboard app and can be effectively used to prevent cyberbullying and keep track of the usage of bad words by any user. The main challenge in this model is to find the perfect balance between the choice of alternative words and the reduction in offense percentage the word provides. In the end, it can be concluded that it is very much possible and most probably the best way to reduce cyberbullying using such models. The accuracy of the model is 97.777%. Besides, the user can just check the offensiveness percentage live right in front of him as and when he types and can choose to remove the offensiveness words altogether from his sentence without much inconvenience. This kind of implementation has a lot of potentials and can be provided as a public API or as an extension to anyone who wants to integrate it into their keyboard.

References

1. 10 Social Media Statistics You Need to Know in 2020 [Infographic]. https://www.oberlo.in/blog/social-media-marketing-statistics#:%CB%9C:text=It%E2%80%99s%20without%20a%20doubt%20one,population%20(Emarsys%2C%202019)
2. 11 facts about cyberbullying. https://www.dosomething.org/us/facts/11-facts-about-cyber-bullying
3. A Majority of Teens Have Experienced Some Form of Cyberbullying. https://www.pewresearch.org/internet/2018/09/27/a-majority-of-teens-have-experienced-some-form-of-cyberbullying/
4. Chuang, Y.T., Lombera, I.M., Melliar-Smith, P.M., Moser, L.E.: Detecting and defending against malicious attacks in the iTrust information retrieval network. In: ICOIN (2012). https://itrust.ece.ucsb.edu/
5. Casey, M.A., Veltkamp, R., Goto, M., Leman, M., Rhodes, C., Slaney, M.: Content-based music information retrieval: current directions and future challenges. Proc. IEEE **96**(4), 668–696 (2008). https://doi.org/10.1109/JPROC.2008.916370
6. David, M.: Mountain spatial filters for mobile information retrieval GIR 2007. In: Proceedings of the 4th ACM workshop on Geographical Information Retrieval, pp. 61–62, November 2007
7. Pennington, J., Socher, R., Manning, C.D.: GloVe: global vectors for word representation. In: Proceedings of the 2014 Conference on Empirical Methods in Natural Language Processing, pp. 1532–1543 (2014)
8. Zhou, J., Xu, W.: End-to-end learning of semantic role labeling using recurrent neural networks. In: Proceedings of the 53rd Annual Meeting of the Association for Computational Linguistics (2015)
9. Hammerton, J.: Named entity recognition with long short-term memory. In: Proceedings of the Seventh Conference on Natural Language Learning (2003)
10. Chiu, J.P., Nichols, E.: Named entity recognition with bidirectional LSTM-CNNs. Trans. Assoc. Comput. Linguist. **4**, 357–370 (2016)
11. Nandhini, B.S., Sheeba, J.I.: Online social network bullying detection using intelligence techniques. Proc. Comput. Sci. **45**, 485–492 (2015)
12. Hosseinmardi, H., Rafiq, R.I., Han, R., Lv, Q., Mishra, S.: Prediction of cyberbullying incidents in a media-based social network. In: 2016 IEEE/ACM International Conference on Advances in Social Networks Analysis and Mining (ASONAM), pp. 186–192, August 2016
13. Huang, Q., Singh, V.K., Atrey, P.K.: Cyber bullying detection using social and textual analysis. In: Proceedings of the 3rd International Workshop on Socially-Aware Multimedia, SAM 2014, pp. 3–6. ACM, New York (2014)

14. Reynolds, K., Kontostathis, A., Edwards, L.: Using machine learning to detect cyberbullying. In: 2011 10th International Conference on Machine Learning and Applications and Workshops, vol. 2, pp. 241–244, December 2011
15. Dinakar, K., Reichart, R., Lieberman, H.: Modeling the detection of textual cyberbullying. In: The Social Mobile Web (2011)
16. How many languages are there in the world? (2016). https://www.ethnologue.com/guides/how-many-languages. Accessed 10 Jan 2019
17. Bartlett, M.S., Littlewort, G.C., Frank, M.G., Lee, K.: Automatic decoding of facial movements reveals deceptive pain expressions. Curr. Biol. **24**(7), 738–743 (2014)
18. Rinn, W.E.: The neuropsychology of facial expression: a review of the neurological and psychological mechanisms for producing facial expressions. Psychol. Bull. **95**(1), 52–77 (1984)
19. Reisenzein, R., Studtmann, M., Horstmann, G.: Coherence between emotion and facial expression: evidence from laboratory experiments. Emot. Rev. **5**(1), 16–23 (2013)
20. Bonanno, G.A., et al.: When the face reveals what words do not: facial expressions of emotion, smiling, and the willingness to disclose childhood sexual abuse. J. Pers. Soc. Psychol. **83**(1), 94 (2002)
21. Cristani, M., Raghavendra, R., Del Bue, A., Murino, V.: Human behavior analysis in video surveillance: a social signal processing perspective. Neurocomputing **100**, 86–97 (2013). Special issue: Behaviours in video
22. Valiune, D. and Perminas, A.: Differences in anger, aggression, bullying among adolescents in different self-esteem groups. Global J. Guidance Counseling Schools: Curr. Perspect. **6**, 61–67 (2017)
23. Bosworth, K., Espelage, D.L., Simon, T.R.: Factors associated with bullying behavior in middle school students. J. Early Adolesc. **19**(3), 341–362 (1999)
24. Wang, X., Yang, L., Yang, J., Wang, P., Lei, L.: Trait anger and cyberbullying among young adults: a moderated mediation model of moral disengagement and moral identity. Comput. Hum. Behav. **73**, 519–526 (2017)
25. Hussian, A., Sharma, S.: Anger expression and mental health of bully perpetrators. FWU J. Soc. Sci. **8**(1), 17 (2014)
26. Dinakar, K., Jones, B., Havasi, C., Lieberman, H., Picard, R.: Common sense reasoning for detection, prevention, and mitigation of cyberbullying. ACM Trans. Interact. Intell. Syst. **2**(3), 1–30 (2012)
27. Shome, A., Rahman, M.M., Chellappan, S., Islam, A.A.A.: A generalized mechanism beyond NLP for real-time detection of cyber abuse through facial expression analytics. In: Proceedings of the 16th EAI International Conference on Mobile and Ubiquitous Systems: Computing, Networking and Services, pp. 348–357 (2019)
28. Hussian, A., Sharma, S.: Anger expression and mental health of bully perpetrators. FWU J. Soc. Sci. **8**(1), 17 (2014)
29. Kaggle Dataset. https://www.kaggle.com/c/jigsaw-toxic-comment-classification-challenge/data. Accessed 25 Jan 2020

An Experimental Study on Exact Multi-constraint Shortest Path Finding

Xuanyi Zhang$^{(\boxtimes)}$, Ziyi Liu, Mengxuan Zhang, and Lei Li

University of Queensland, Brisbane, Australia
{xuanyi.zhang1,ziyi.liu3,mengxuan.zhang}@uq.net.au,
l.li3@uq.edu.au

Abstract. Given a directed graph with each edge has a weight and several criteria, a multi-constraint shortest path query (*CSP*) asks for the shortest path that satisfies the constraints on these criteria. It is a general routing problem and can be used to customise user's routing requirements. However, only the single-constraint version has been well studied while the multi-constraint version is ignored due to its complexity. In this paper, we explore the multi-*CSP* problem by extending three existing single-*CSP* algorithms (*Skyline-Dijkstra*, *sKSP*, and *eKSP*) and experimentally study their performance and behaviours. Experiment results provides insights of how the query distance, constraint ratio, criteria number, strictness, and correlation influence the query performance.

Keywords: Constraint shortest path · Road network

1 Introduction

With the advance of the transportation-related smart city technologies, we have our road network information more detailed than ever before. For instance, apart from the ordinary shortest path, we now have more road information such as the traveling time [1,2], toll charge, number of big turns, increase of elevation, number of traffic lights, fuel consumption, battery usage, tourist interests, number of humps, height/weight limits, and etc. By combining these criteria interchangeably, we can provide customised trip that satisfies user's flexible needs. In other words, this *multi-criteria* route planning is a more generalised way to plan our trips, while the ordinary *shortest path* is a special case. However, when we have more than one optimisation goals, it is very likely we cannot find one result that is optimal in every criterion. In fact, we normally use a set of *skyline* results [3] in a multi-objective problem like this. Therefore, one way to implement it is through *skyline path* [4–6], which extends the skyline concept to path finding.

However, the *skyline path* is normally not a good choice because the following reasons. Firstly, it tends to generate a large number of results for users to choose, with some results are very good at some criteria but poor at others. For example, suppose we have two criteria: distance and toll charge. Then the two extreme paths that one has the lowest toll but has to take a long detour, while the other one has the shortest distance but has a high toll charge, all belong to the

© Springer Nature Switzerland AG 2021
M. Qiao et al. (Eds.): ADC 2021, LNCS 12610, pp. 166–179, 2021.
https://doi.org/10.1007/978-3-030-69377-0_14

skyline path results. However, both of them are very unlikely be a satisfactory result, so it is a waste of time to compute them. Secondly, it is very time-consuming to compute the skyline paths, as its time complexity is $O(w_{max}|V| \times (|V|\log|V| + |E|))$ ($|V|$ is the vertex number, $|E|$ is the edge number, and c_{max} is the largest edge cost, suppose cost is smaller than weight)) [7]. Therefore, the *Constraint Shortest Path (CSP)* is a more practical solution, which only optimises on one criterion while requires the other criterion satisfying their pre-defined thresholds (constraints). In this way, only the best path that satisfies all the user requirements is returned.

Therefore, in this work, we extend the existing single-*CSP* algorithms to the multi-criteria environment to test their performance and behaviours. Like the ordinary shortest path algorithms, the *CSP* algorithms can also be classified into *index-free* and *index-based* methods. We only focus on the *index-free* methods because they provide the ground-truth results of the multi-*CSP* queries and extending them is the first step towards exploring the multi-*CSP* problem. Besides, it is unclear how the *index* can be utilised flexibly for different numbers of query constraints because the criteria number has to be fixed. We will leave the index-based multi-*CSP* to the future work. In terms of the *index-free* methods, there are two main streams: 1) *Dijkstra*'s-like expansion, which expands the search space incrementally and prune the dominated sub-paths until reaching the destination, and 2) k-shortest path enumeration, which tests the paths in the distance-increasing order until obtaining the first valid result. To test their performance in multi-criteria, we extend the algorithms from both streams. Specifically, we extend the following three algorithms: *Skyline Dijkstra* [7], *Search-based kSP* [8], and *Enhanced kSP* [9]. Finally, we conduct extensive experiment on the influence of query distance, constraint ratio, criteria number, stricter criteria, and criteria correlation with the weight, and provide insightful analysis. These experiment results provides insights of how the problem evolves and how the algorithms perform in the higher dimensional space. Our contributions are listed below:

- We formally study the multi-*CSP* problem, which is the most general version of *CSP*;
- We extend three existing single-*CSP* algorithms to the multi-*CSP* environment efficiently;
- Extensive experiments on the extended algorithms fully explore and provide insights of their performance under different scenarios.

The rest of this paper is organised as follows: We first discuss the related works in Sect. 2. Then we define the problem formally in Sect. 3 and describe three multi-*CSP* algorithms in Sect. 4. Experimental study is presented in Sect. 5, and Sect. 6 concludes the paper.

2   Related Work

In this section, we introduce the existing solutions for multi-constraint shortest path finding which include *Skyline Path*, *Single-CSP* and *Multi-CSP*.

2.1   Skyline Path

Skyline-Dijkstra's[1] [7] is the first algorithm for skyline path query that expands the search space incrementally and dominance relation is applied to prune the search space. In this algorithm, one vertex might be visited multiple time with each search corresponding to one skyline result. Then [4] incorporates *landmark* [10] to estimate the lower bounds and prunes the search space with dominance relations. [5] utilizes the skyline operator to find a set of skyline destinations.

2.2   Single-CSP

Exact Algorithms. [11] first studies the *CSP* by applying the dynamic programming algorithm. After that, [12] views the *CSP* as an *integer linear programming (ILP)* problem and uses the standard *ILP* algorithm to answer it. However, both of them only work on small toy graphs and can hardly scale to the real-life road networks containing hundreds of thousands of vertices. Among the practical solutions, the *Skyline Dijkstra* algorithm sets a constraint on the cost during the skyline search and the search space can be further pruned as the *CSP* result is a subset of the skyline paths. *k-Path* enumeration [8,9,13] is another practical solution, which tests the top-k paths in the distance-increasing order one by one until obtaining the first one that satisfies the constraint. Specifically, *Pulse* [13] improves this procedure by applying the constraint pruning, but its performance is limited because it is based on depth-first search. *sKSP* [8] generates the k-paths faster with the help of a reverse shortest path tree. Finally, *eKSP* [9] further replaces *sKSP*'s search with sub-path concatenation to speed up k-path generation. Although many shortest path indexes [14–17] are widely used for efficient query processing, only *CH* [18] has been extended to *CSP* [19]. This is because the query constraints are not fixed with the constraint value among a range, such that the index has to capture all the possible solutions and its shortcuts are actually *skyline paths*. As a result, it is *Skyline-Dijkstra*-based with long construction time, inefficient query processing and huge index size. In fact, *CSP-CH* has to reduce the skyline paths number in each shortcut to reduce the construction time, but it would deteriorate the query performance.

Approximate Algorithms. The approximate *CSP* finds the paths with length smaller than $(1 + \alpha)$ times of the optimal path. [7] is the first approximate method with a complexity of $O(|E|^2 \frac{|V|^2}{\alpha - 1} \log \frac{|V|}{\alpha - 1})$, and [20] reduces it to $O(|V||E|(\log\log \frac{w_{max}}{w_{min}} + \frac{1}{\alpha - 1}))$. *Lagrange Relaxation* obtains the approximation result [21] with $O(|E| \log^3 |E|)$ iterations of path finding. However, these approximate algorithms are several orders of magnitude slower than the *k-Path*-based exact algorithms as tested in [22] and [13]. *CP-CSP* [23] approximates the *Skyline*

---

[1] It was called *MinSum-MinSum* problem as the term *skyline* was not used in the 1980s.

Dijkstra by distributing the approximation power $\sqrt[n]{\alpha}$ to the edges. Nevertheless, its approximation ratio would decrease to 1 when the path is long, so its performance is no better than *Sky-Dijk* in practice. The only practical approximate index method is *COLA* [24], which partitions the graph into regions and precomputes approximate skyline paths between regions. Similar to *CP-CSP*, the approximate paths are computed in the *Skyline-Dijkstra* fashion. In order to avoid the diminishing approximation power $\sqrt[n]{|V|}$, it views α as a budget and concentrates it on the important vertices. Similar pruning technique is also applied in the time-dependent path finding [25]. In this way, the approximation power is released so it has faster construction time and smaller index size.

2.3 Multi-CSP

The Multi-*CSP* algorithms can be classified into *scalar* and *Pareto* methods. Scalar methods transform the question to a single-objective problem based on existing heuristics [26] such as a weighted objective function [27]. However, such methods are not easy to achieve coherence on a set of criteria [28]. On the other sides, the *Pareto* [29–31] method uses the idea of *Pareto dominance*. The exact ways to find the entire set of *Pareto*-optimal paths, or called skyline paths, can be divided into labeling and ranking algorithms. Ranking algorithms produce a set of the k-shortest path first and then discard the paths dominated by other criteria in the path set. Note that the fundamental solutions on solving MCSP are similar to the methods on Bi-Criteria Path Problem, which is a specific case of multi-*CSP*. It will be an easy extension for solving multi-*CSP* on a given pair of source and destination. However, there are much more challenging on answering an arbitrary OD pair, as the skyline paths will increase dramatically when the number of criteria increases.

3 Preliminary

3.1 Problem Definition

Let $G(V, E)$ be a graph with vertex set V and edge set $E \subseteq V \times V$. Each edge $e \in E$ has d dimensional values, including a weight $w(e)$ and a set of costs $\{c_1(e), \ldots, c_{d-1}(e)\}$. By considering w and costs together, we have d criteria in total. A path p from the source $s \in V$ to the target $t \in V$ is a sequence of consecutive vertices $p = \langle s = v_0, v_1, \ldots, v_k = t \rangle$, where $(v_n, v_{n+1}) \in E, \forall n \in [0, k-1]$. Each path p has a weight $w(p) = \sum_{e \in p} w(e)$ and a set of costs $\{c_i(p) = \sum_{e \in p} c_i(e) | i \in [1, d-1]\}$. Then we define the multi-constrained shortest path query below:

Definition 1 (Multi-CSP Query). *Given a graph G, a source s, a target t, and a set of maximum cost bounds $\overline{C} = \{C_1, ..., C_{d-1}\}, \forall C_i \in \mathbb{R}^+$, where $i \in [1, d-1]$. A Multi-CSP query $q(s, t, \overline{C})$ returns a path with the minimum $w(p)$ while guaranteeing that each $c_i(p) \leq C_i$.*

3.2 Skyline Path

It should be noted that a Multi-CSP path from s to t is one of the skyline paths from s to t. Besides, our algorithms and pruning techniques are also based on the skyline path, so we define the skyline path here. Like the other skyline problems, the skyline path is also based on the notion of *dominance*, which is defined below:

Definition 2 (Path Dominance). *Given two paths p_1 and p_2 with the same OD, p_1 dominates p_2 iff $w(p_1) \leq w(p_2)$ and $c_i(p_1) \leq c_i(p_2), \forall i \in [1, d-1]$, and at least one of them is the strictly smaller relation.*

Definition 3 (Skyline Path). *Given any two paths p_1 and p_2 with the same OD, they are skyline paths if they cannot dominate each other.*

Then the skyline path set between any s and t is the minimal set of paths that cannot dominate each other, but can dominate all the other paths. Obviously, *CSP* path is a skyline path, because otherwise there would another path with smaller weight and also satisfying all the constraints. And our *CSP* path is the shortest one among all the constraint-satisfying skyline paths.

4 Multi-CSP Algorithms

In this section, we extend three *CSP* algorithms to the multi-*CSP* scenario. Specifically, *Skyline-Dijkstra (Sky-Dij)* is the basis whose searches replace the classic shortest path expansion with skyline path expansion, *Search-based kPath (sKSP)* enumerates the shortest paths in the distance-increasing order with the help of *Sky-Dij*, and *Enhanced kPath (eKSP)* further avoids expensive *Sky-Dijk* to achieve better performance.

4.1 Skyline-Dijkstra's Multi-CSP

The Skyline-Dijkstra's algorithm finds all the skyline paths from s to t in the same manner of the *Dijkstra's*. It can be used to answer *CSP* query by pruning with the constraint. Different from the *Dijkstra's* that computes the shortest distance from s to each visited vertex in the distance-increasing order, the *Skyline-Dijkstra* stores a set of skyline paths at each visited vertex by assigning a label set $L(v)$ to each vertex $v \in V$. $L(v)$ is empty in the beginning and then maintained as the current set of skyline paths from s to v. Moreover, the vertex order in the priority queue Q in *Skyline-Dijkstra* depends on multiple criteria including a weight and a set of costs, which is unlike the Q in *Dijkstra's* sorting on the shortest distance. Therefore, $L(v)$ in Q has no strict order and we can select either weight or one of the costs as its ranking order. In this work, we still maintains Q in ascending order of each vertex's weight.

For a multi-*CSP* query $q(s, t, \overline{C})$, the primary procedure of a traversal is shown as follows:

1. Initially, Q includes path p_0 from s to s.
2. In each iteration, *Skyline-Dijkstra* pops the top path p with the minimum weight from Q. Suppose v is the top vertex in p.
3. Then we check whether v is equal to the destination t. If $v \neq t$, it means that p has not reached the destination. Then we expand p to a path p', which can be obtained by adding an edge (v, v') at the end of p.
4. Next, we check the feasibility of p'. If any $c_i(p') > C_i$ or p' dominated by other paths in $L(v')$, we can discard p' safely. Otherwise, we add p' to Q and also insert it to $L(v')$. Meanwhile, we validate the paths in $L(v')$ to eliminate the dominated paths.
5. The algorithms terminates when Q is empty or t is reached. Then the path with minimum weight in the label set $L(t)$ is the result.

It should be noted that Q has to maintain a set of skyline paths from s to each visited vertices. On the computing of skylines, the number of skyline paths in Q will increase dramatically, which is caused by the following reasons: 1) The *Dijkstra*-based search presents circle space, so the number of visited vertices will be exploding when the query distance is long. 2) The *Skyline-Dijkstra* may visit the same vertex more than once to produce different skyline paths. 3) When the dimensions number increases, more skyline paths are generated as they become harder to dominate each other. Hence, *Skyline-Dijkstra* may spend hours to answer a multi-*CSP* query in the cases of long query distance and high dimensions.

Pruning Technique 1. To improve the query performance of the *Skyline-Dijkstra's*, we propose the following pruning techniques. It should be noted that if a path p is a skyline path, the sub-paths of p are all skyline paths [19]. Then in the Step 4 above, suppose path p' is identified as a current skyline path from s to v', and it is added to Q and $L(v')$. However, it could be dominated by the paths that are found later. Therefore, if p' is not a sub-path of a skyline path from s to t, it is unnecessary to further check its neighbours when it is popped out from Q. In the pruning, for each vertex v popped out from Q, we will identify the dominate relations of its corresponding path p with the paths in $L(v)$. If p is dominated, we will stop the exploring on the path and pop out the next top vertex in Q. This process will reduce the size of the priority queue and avoid the search on the unnecessary vertices.

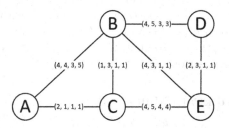

Fig. 1. Example of Skyline-Dijkstra's Multi-*CSP*

Running Example. An example road network is shown in Fig. 1, which has four criteria on each edge. We consider the first criterion as weight and other three as costs. Suppose we have a multi-CSP query $q(A, E, \overline{C})$, where $\overline{C} = \{8, 9, 10\}$. The *Skyline-Dijkstra's* initializes Q with a A to A path p_0, which includes zero for all criteria. Then, $p_0 = \langle A, A \rangle$ is popped out. By checking the neighbours of A, we can get a path $p_1 = \langle A, B \rangle$ with $w(p_1) = 4$ and $c(p_1) = \{4, 3, 5\}$, and another path $p_2 = \langle A, C \rangle$ with $w(p_2) = 2$ and $c(p_2) = \{1, 1, 1\}$. We assign p_1 to $L(B)$ and p_2 to $L(C)$. In consequence, we push p_1 and p_2 into Q. In the next round, p_2 is popped out from Q as $w(p_2)$ is the minimum. We generate path $p_3 = \langle A, C, B \rangle$ with $w(p_3) = 3$ and $c(p_3) = \{4, 2, 2\}$, and path $p_4 = \langle A, C, E \rangle$ with $w(p_4) = 6$ and $c(p_4) = \{6, 5, 5\}$. We maintain the labels by adding p_3 to $L(B)$ and p_4 to $L(E)$. Because $L(B)$ already has $p_1 = \{4, 4, 3, 5\}$ and p_1 is dominated by p_3, we delete p_1 from $L(B)$. Furthermore, We only push p_3 into Q as p_4 has already reached E. As the current path with minimum weight is p_3, we pop it out and obtains $p_5 = \langle A, C, B, D \rangle$ with $w(p_5) = 7$ and $c(p_5) = \{9, 5, 5\}$, and $p_6 = \langle A, C, B, E \rangle$ with $w(p_6) = 7$ and $c(p_6) = \{7, 3, 3\}$. Note that the first cost c_1 of p_5 is 9, which exceeds the corresponding constraint $C_1 = 8$. Thus, we stop expanding from path p_5 in the future. For p_6, as it cannot be dominated by p_4, we push it in $L(B)$ as a skyline path. Currently, Q still has $p_1 = \{4, 4, 3, 5\}$. When we pop it out, we notice that it has already been dominated by the current path in $L(B)$. Therefore, we stop checking the neighbours of B. By checking the next top element in Q, we find that it is empty, and algorithm terminates.

Complexity Analysis. The time complexity of the *Skyline-Dijkstra's* in the two dimensional space is $O(c_{max}|V|(|V|log|V| + |E|))$, where c_{max} is the largest edge cost and suppose costs are smaller than weights. This is because c_{max} is the worst case scenario of the skyline number. When it comes to the multi-criteria, this worst case explodes to $\prod c_{i \cdot max}$), and the time complexity becomes $O(\prod c_{i \cdot max}|V|(|V|log|V| + |E|))$. Therefore, the *Skyline-Dijstra's* performance deteriorates dramatically as the dimension number increases.

Pruning Technique 2. To further improve the query performance by reducing the skyline path numbers, we first compute a set of minimum cost to t. Specifically, for each cost, we run a reverse *Dijkstra's* on it and obtain the minimum cost from each vertex to t. Then during the *Skyline-Dijkstra's* search, when we expand a path p', we can compute the lower-bounds of the costs of p' by adding the current cost with the pre-computed minimum cost. If any of the cost lower bound exceeds its constraint, we can prune p' directly as it cannot produce a valid result. Moreover, the reversed *Dijkstra's* can prune with the constraints and leaving the unvisited costs ∞. Because the *Dijkstra's* complexity is smaller than the *Skyline-Dijkstra's*, the pre-computation won't affect the time complexity, but it can boost the query performance in practice.

4.2   Search-Based kPath Multi-CSP

The *Skyline-Dijkstra's* has to find all the constraint-satisfying skyline sub-paths by expanding the search space incrementally from s before the result is found.

However, it is time-consuming even if the shortest path itself is the multi-CSP result. Therefore, the $kPath$ enumeration [32], which finds the top-k shortest paths in the distance-increasing order, can work in this scenario perfectly. By testing the shortest paths strictly in the distance-increasing order, the final result is the first one that satisfying all the constraints. The $kPath$ algorithm first find the shortest path p_1. To find the second shortest path, it first deviates the from the vertices v_i on p_1 iteratively and finds the shortest path from v_i to t that does not traverse any vertex in the prefix subpath $p_1(s \rightarrow v_i)$. Then a set of candidate paths are obtained by concatenating $p_1(s \rightarrow v_i)$ and the shortest path from v_i to t. It is easy to prove that the second shortest path is among these candidates. To this end, we put these candidates into a priority queue Q, and use the one on top as the second one, and further expands from it to generate and insert another set of candidates into Q. It should be noted that only the subpath from v_i to t needs deviation as the previous prefix $s \rightarrow t$ is also covered by the previous deviations. To further speedup the enumeration, a reverse shortest path tree can be utilised to speed up the deviation search [8].

However, enumerating all the shortest paths until reaching the final result could also be time-consuming as there could be hundreds of thousands of candidates to test. In fact, the $kPath$ algorithm has a time complexity of $O(k|V|(|V|\log|V|+|E|))$. Then it would be a disaster if k is large for a multi-CSP result. Therefore, the above two pruning techniques and the skyline dominance are also introduced into the $kPath$ computation to reduce the candidate number and the enumeration number k. Specifically, each vertex also has a skyline set $L(v)$ to prune the dominated candidates. When we conduct a deviation search, its search space can be pruned with the same procedure as the *Skyline-Dijkstra's*. In fact, we can view this search-based $kPath$ algorithm as a set of tiny *Skyline-Dijkstra's* from each deviation vertex. Therefore, its performance is also limited by the *Skyline-Dijkstra's*.

4.3   Enhanced kPath Multi-CSP

To further improve the query performance, we can get rid of *Skyline-Dijkstra's* algorithm's *Dijkstra's* part completely and only utilise the *Skyline* part in the $kPath$ enumeration. Besides, we can also reduce the candidate number by avoiding generating the complete candidate set for each path and only generate one when necessary. Therefore, instead of generating all the deviation paths from the current path, we first compute the *deviation cost* of all the deviation edges. For example, suppose u is on p_i and v is a neighbor of u but not on p_i, then edge (u, v) is a deviation edge and its deviation cost is $w(u, v) + d(v, t) - d(u, t)$. By organizing all the deviation edges in a heap, we can generate new paths in the distance-increasing order. However, there is no need to generate them all at the same time as it would be the same as the *search-based kPath*. During the enumeration, we only need to generate one candidate from the top path, and another candidate from its parent path, i.e., the path that generated the current top path. This is because we can view the $kPath$ search space divided by the paths and deviation edges into subspaces, and we only need to guarantee each subspace has an active candidate to consider in Q. Therefore, when we process the top path p', only the old subspace that generates p' and the new subspace that p' creates have no candidate. The first case can be fixed by generating the next shortest path from p''s parent path, and the second case is taken care by p'. In this way, only two candidates instead of $|p'|$ candidates are generated in each iteration, so we can avoid a large amount of expensive *Skyline-Dijkstra's* searches.

Similar pruning techniques can also be applied in this algorithm. When we generate a new candidate path by deviating to a neighbor v, we can first test it with v's skyline paths $L(v)$ and either prune it or prune the skyline paths. Then we can also prune with the feasibility when any cost does no satisfy its constraint. If any of the above two conditions hold, we can discard this candidate safely and generate the next one until all the constraints are satisfied.

5   Experiments

5.1   Experimental Settings

We implement all the algorithms in C++ with full optimizations. The experiments are conducted in a 64-bit Ubuntu 18.04.3 LTS with two 8-cores Intel Xeon CPU E5-2.9 GHz and 186 GB RAM.

Datasets. All the tests are conducted in the real New York network obtained from the 9th DIMACS Implementation Challenge[2]. It has 264,346 vertices and 733,846 edges. Each edge has a distance and a travel time, and we treat the distance as weight and the travel time as cost.

Query Set. We use five sets of OD pairs Q_1 to Q_5 (100 each) as the base query set, and each of them represents a distance category. Specifically, we first estimate the network diameter w_{max} with an approximation method [33], then each Q_i are the random OD pairs with distance within $[w_{max}/2^{6-i}, w_{max}/2^{5-i}]$. For example, Q_1 are the nearer OD pairs within $[w_{max}/32, w_{max}/16]$ and Q_5 are the far away OD pairs within $[w_{max}/2, w_{max}]$. We run each query five times and report the average time for the 100 queries in each query set.

Then we generate the costs of the edges. Specifically, there are three types of cost:

1. *Positive Correlation*: The cost grows proportionally to the distance such travel time and fuel consumption, and we use the travel time in the dataset directly in the single-criteria test.
2. *Random*: The cost has no clear relation with distance like the slope gradient, and we generate it randomly in the multi-criteria test.
3. *Negative Correlation*: The cost drops proportionally to the distance that causes the worst case in the skyline operations. We use the equation $c = \frac{k}{w}, k \in \mathbb{R}^+$ to generate cost for each edge.

---

[2] http://users.diag.uniroma1.it/challenge9/download.shtml.

For the multi-criteria tests, we set the criteria number $|m|$ as 1, 2, 3, 4, 5, 10, 15, and 20. To set the query constraint C_i to these OD pairs, we first compute the minimum cost C^i_{min} of criteria m_i such that if $C_i < C^i_{min}$, the optimal result can not be found. Then we also find the maximum cost C^i_{max} of criteria m_i such that if $C_i > C^i_{max}$, the optimal results is the path with the shortest distance. In this way, we obtain a cost range $[C^i_{min}, C^i_{max}]$. Furthermore, in order to test the constraint influence on the query performance, we generate five constraints for each OD pair: $C_i = r \times C^i_{max} + (1 - r) \times C^i_{min}$, with $r = 0.1, 0.3, 0.5, 0.7$ and 0.9. When r is small, C is closer to the minimum cost; when r is big, C is closer to the maximum cost.

5.2 Query Distance

Figure 2 shows the query processing time of different algorithms in NY under different constraint ratios r in the distance-increasing order. For starter, all the algorithms' running time increase as the distance increases. When the ODs are near to each other like Q_1 and Q_2, all of them can finish in 10 s. When the distance further increases, they soar up to thousands of seconds. Specifically, *Sky-Dij* is the fastest when the distance is short but its performance deteriorates for longer queries because it has a huge amount of intermediate partial skyline results to extend. The two *KSP* methods perform better when the distance becomes longer, especially when r is larger. This is because they enumerate the paths in the distance-increasing order, and the larger r tends to terminate the enumeration earlier. Among the two *KSP* methods, *eKSP* is faster because it avoids the expensive *Dijkstra's* search during the enumeration.

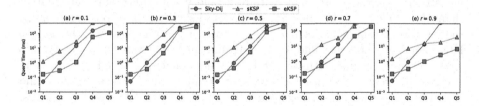

Fig. 2. Query distance

5.3 Constraint Ratio

Figure 3 shows the query processing time from the perspective of r. The *Sky-Dij* and *KSP* methods have different trends: *Sky-Dij* grows as the r increases, while *cKSP* and *eKSP* have a hump shape. It should be noted that because the time is shown in logarithm, a very small change over 10^2 in the plot actually has a very large difference. These trends become more obvious when the query distance becomes larger (Q_3 to Q_5). The reason of the increase trend is that *Sky-Dij* expands the search space incrementally, during which they accumulate a set of

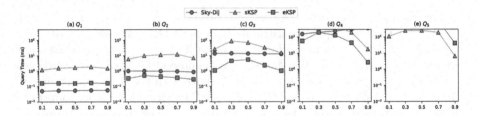

Fig. 3. Constraint ratio

partial skyline paths. Therefore, when the r is small, partial skyline path set is also small because most of the infeasible ones are dominated, so it has faster running time. When r is big, the partial skyline path set is also large so it runs slower. As for the hump trend, the two *KSP* algorithms test the complete paths iteratively in the distance-increasing order. Therefore, when r is big, more paths would satisfy this constraint, so there is a higher chance for the shorter paths to become the final result, and we only need to enumerate a small number of path to find it.

5.4   Number of Constraints

Because Q_1 is too short to tell the difference and Q_3 is slow to run, we use Q_2 to test the influence of criteria number m. The results are shown in Fig. 4 (a)–(c). Because not all the query constraints can be satisfied, we also show the *invalid query number* with bar. Firstly, as the criteria number increases, the invalid query number also increases. Specifically, $r = 0.1$ has more invalid queries because it has stricter criteria (more than half are invalid when criteria number is larger than 15), while $r = 0.9$ has fewer because its criteria are looser. Secondly, the query answering time increases as criteria number increases. Although the larger number of criteria could prune the result space by invalid sub-paths, it also generates more skyline paths as the high dimensional space tend to have more skylines. Therefore, the pruning power of skyline paths becomes weaker, and it is also more time-consuming to test and maintain the skyline path results. Consequently, *Sky-Dij*'s running time soars up, and only finish query in time when r is small. In addition, *sKSP*'s performance is also affected by it skyline search. Finally, *eKSP* is the fastest as it avoids the expensive the *Sky-Dij*, while the two search-based methods are much slower.

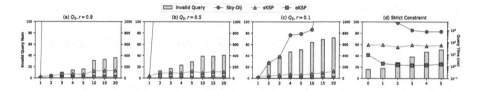

Fig. 4. Constraint number. Line: query time. Bar: invalid query number

5.5   Influence of Stricter Constraints

The previous experiments have the same r for each query. In this experiment, we test the influence of different r in the same query. Specifically, we set $r = 0.9$ and $m = 5$ as default, and increases the number of $r = 0.1$ from 0 to 5. The results are shown in Fig. 4-(d). For starter, as the number of $r = 0.1$ increases, more queries become invalid. Consequently, more sub-queries would become invalid, so the query time of $eKSP$ and Sky-Dij drop. As for the $sKSP$, its query time does not have a clear trend. As the $eKSP$ is the fastest method, we can draw the conclusion that stricter criteria could improve the query performance, while looser criteria would prolong the query time.

5.6   Constraint Correlation

In this experiment, we also compare the query performance when the criteria are random or negative correlation with the weight, and the result is shown in Fig. 5. First of all, the positive correlation is the fastest to run. This is because it generates fewer skyline paths and has smaller result space. Secondly, random correlation is slower, and it is the most common type of criterion in real-life. Finally, the negative correlation is much slower, because almost all the paths are skyline paths, which booms the search space. Therefore, even if the criteria number is only one, its performance deteriorates dramatically such that some of the tests cannot finish. Fortunately, although the negative correlation is worst-case scenario of skyline and CSP path, it is very rare in real-life.

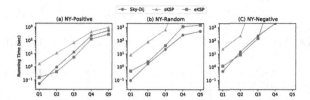

Fig. 5. Correlation influence

6 Conclusion

In this paper, we study the multi-constraint shortest path problem as it is more general and more powerful to satisfy users' needs. Specifically, we extend three state-of-the-art *CSP* algorithms (*Sky-Dij*, *sKSP*, and *eKSP*) to the multi-criteria scenario. After that, we conduct extensive experiments to test influence of distance, constraint ratio, stricter criteria, and criteria correlation with weight. These experimental results provide insights of how the multi-*CSP* is different from *CSP*, and guide future research for more efficient multi-*CSP* query processing.

References

1. Li, L., Hua, W., Du, X., Zhou, X.: Minimal on-road time route scheduling on time-dependent graphs. PVLDB **10**(11), 1274–1285 (2017)
2. Li, L., Zheng, K., Wang, S., Hua, W., Zhou, X.: Go slow to go fast: minimal on-road time route scheduling with parking facilities using historical trajectory. VLDB J. **27**(3), 321–345 (2018)
3. Papadias, D., Tao, Y., Fu, G., Seeger, B.: An optimal and progressive algorithm for skyline queries. In: SIGMOD, pp. 467–47 (2003)
4. Kriegel, H.-P., Renz, M., Schubert, M.: Route skyline queries: a multi-preference path planning approach. In: ICDE 2010, pp. 261–272. IEEE (2010)
5. Gong, Q., Cao, H., Nagarkar, P.: Skyline queries constrained by multi-cost transportation networks. In: ICDE, pp. 926–937. IEEE (2019)
6. Ouyang, D., Yuan, L., Zhang, F., Qin, L., Lin, X.: Towards efficient path skyline computation in bicriteria networks. In: Pei, J., Manolopoulos, Y., Sadiq, S., Li, J. (eds.) DASFAA 2018. LNCS, vol. 10827, pp. 239–254. Springer, Cham (2018). https://doi.org/10.1007/978-3-319-91452-7_16
7. Hansen, P.: Bicriterion path problems. In: Multiple Criteria Decision Making Theory and Application, pp. 109–127. Springer (1980). https://doi.org/10.1007/978-3-642-48782-8_9
8. Gao, J., Qiu, H., Jiang, X., Wang, T., Yang, D.: Fast top-k simple shortest paths discovery in graphs. In: CIKM, pp. 509–518 (2010)
9. Sedeño-Noda, A., Alonso-Rodríguez, S.: An enhanced K-SP algorithm with pruning strategies to solve the constrained shortest path problem. Appl. Math. Comput. **265**, 602-618 (2015)
10. Kriegel, H.-P., Kröger, P., Kunath, P., Renz, M., Schmidt, T.: Proximity queries in large traffic networks. In: ACM GIS, pp. 1–8 (2007)
11. Joksch, H.C.: The shortest route problem with constraints. J. Math. Anal. Appl. **14**(2), 191–197 (1966)
12. Handler, G.Y., Zang, I.: A dual algorithm for the constrained shortest path problem. Networks **10**(4), 293–309 (1980)
13. Lozano, L., Medaglia, A.L.: On an exact method for the constrained shortest path problem. Comput. Oper. Res. **40**(1), 378–384 (2013)
14. Zhang, M., Li, L., Hua, W., Zhou, X.: Efficient 2-hop labeling maintenance in dynamic small-world networks. In: ICDE. IEEE (2021)
15. Li, L., Wang, S., Zhou, X.: Fastest path query answering using time-dependent hop-labeling in road network. In: TKDE (2020)

16. Li, L., Zhang, M., Hua, W., Zhou, X.: Fast query decomposition for batch shortest path processing in road networks. In: ICDE (2020)

17. Zhang, M., Li, L., Hua, W., Zhou, X.: Stream processing of shortest path query in dynamic road networks. In: TKDE (2020)

18. Geisberger, R., Sanders, P., Schultes, D., Delling, D.: Contraction hierarchies: faster and simpler hierarchical routing in road networks. In: McGeoch, C.C. (ed.) WEA 2008. LNCS, vol. 5038, pp. 319–333. Springer, Heidelberg (2008). https://doi.org/10.1007/978-3-540-68552-4_24

19. Storandt, S.: Route planning for bicycles–exact constrained shortest paths made practical via contraction hierarchy. In: ICAPS (2012)

20. Lorenz, D.H., Raz, D.: A simple efficient approximation scheme for the restricted shortest path problem. Oper. Res. Lett. **28**(5), 213–219 (2001)

21. Juttner, A., Szviatovski, B., Mécs, I., Rajkó, Z.: Lagrange relaxation based method for the QoS routing problem. In: INFOCOM, vol. 2, pp. 859–868. IEEE (2001)

22. Kuipers, F., Orda, A., Raz, D., Van Mieghem, P.: A comparison of exact and ε-approximation algorithms for constrained routing. In: Boavida, F., Plagemann, T., Stiller, B., Westphal, C., Monteiro, E. (eds.) NETWORKING 2006. LNCS, vol. 3976, pp. 197–208. Springer, Heidelberg (2006). https://doi.org/10.1007/11753810_17

23. Tsaggouris, G., Zaroliagis, C.: Multiobjective optimization: improved FPTAS for shortest paths and non-linear objectives with applications. Theory Comput. Syst. **45**(1), 162–186 (2009). https://doi.org/10.1007/s00224-007-9096-4

24. Wang, S., Xiao, X., Yang, Y., Lin, W.: Effective indexing for approximate constrained shortest path queries on large road networks. PVLDB **10**(2), 61–72 (2016)

25. Li, L., Wang, S., Zhou, X.: Time-dependent hop labeling on road network. In: ICDE, pp. 902–913, April 2019

26. Jozefowiez, N., Semet, F., Talbi, E.-G.: Multi-objective vehicle routing problems. Eur. J. Oper. Res. **189**(2), 293–309 (2008)

27. Shi, N., Zhou, S., Wang, F., Tao, Y., Liu, L.: The multi-criteria constrained shortest path problem. Transp. Res. Part E: Logist. Transp. Rev. **101**, 13–29 (2017)

28. Braekers, K., Caris, A., Janssens, G.K.: Bi-objective optimization of drayage operations in the service area of intermodal terminals. Transp. Res. Part E: Logist. Transp. Rev. **65**, 50–69 (2014)

29. Gandibleux, X., Beugnies, F., Randriamasy, S.: Martins' algorithm revisited for multi-objective shortest path problems with a maxmin cost function. 4OR **4**(1), 47–59 (2006). https://doi.org/10.1007/s10288-005-0074-x

30. Guerriero, F., Musmanno, R.: Label correcting methods to solve multicriteria shortest path problems. J. Optim. Theory Appl. **111**(3), 589–613 (2001)

31. Skriver, A.J., Andersen, K.A.: A label correcting approach for solving bicriterion shortest-path problems. Comput. Oper. Res. **27**(6), 507–524 (2000)

32. Yen, J.Y.: Finding the k shortest loopless paths in a network. Manage. Sci. **17**(11), 712–716 (1971)

33. Meyer, U., Sanders, P.: δ-stepping: a parallelizable shortest path algorithm. J. Algorithms **49**(1), 114–152 (2003)

The Effect of Regional Economic Clusters on Housing Price

Jiaying Kou[1]([✉]), Jiahua Du[1], Xiaoming Fu[2], Geordie Z. Zhang[3], Hua Wang[1], and Yanchun Zhang[1]

[1] Institute for Sustainable Industries and Liveable Cities, Victoria University, Melbourne, Australia
jiaying.kou@live.vu.edu.au
[2] Institute of Computer Science, University of Göttingen, Göttingen, Germany
[3] Melbourne Data Analytics Platform, The University of Melbourne, Melbourne, Australia

Abstract. A good location goes beyond the direct benefits from its neighbourhood. Unlike most previous statistical and machine learning based housing appraisal research, which limit their investigations to neighbourhoods within 1 km radius of the house, we expand the investigation beyond the local neighbourhood and to the whole metropolitan area, by introducing the connection to significant influential economic nodes, which we term *Regional Economic Clusters*. By consolidating with other influencing factors, we build a housing appraisal model, named HNED, including housing features, neighbourhood factors, regional economic clusters and demographic characteristics. Specifically, we introduce regional economic clusters within the metropolitan range into the housing appraisal model, such as the connection to CBD, workplace, or the convenience and quality of big shopping malls and university clusters. When used with the gradient boosting algorithm XGBoost to perform housing price appraisal, HNED reached 0.88 in R^2. In addition, we found that the feature vector from Regional Economic Clusters alone reached 0.63 in R^2, significantly higher than all traditional features.

Keywords: Real estate economics · Regional economic cluster · Housing price prediction · Knowledge discovery · Data mining

1 Introduction

The housing market has a strong impact on the economy worldwide. At the national level, housing-related industries contribute to 15–18% of the Gross Domestic Product (GDP) in the United States, nearly 15% of the GDP in the European Union, and 13% of the GDP in Australia[1]. These numbers show that housing is a leading indicator of a nation's economic cycle [19] and important for

---

[1] https://www.nahb.org/News-and-Economics/Housing-Economics/Housings-Economic-Impact/Housings-Contribution-to-Gross-Domestic-Product.

© Springer Nature Switzerland AG 2021
M. Qiao et al. (Eds.): ADC 2021, LNCS 12610, pp. 180–191, 2021.
https://doi.org/10.1007/978-3-030-69377-0_15

economic and financial stability and growth [2]. Real estate is the biggest assets for most households and also one of the strongest factors as a financial assurance to afford a comfortable retirement life. In 2017, the median net wealth[2] of property-owning households with at least one of the occupants over 65 years old was over \$934,900, whereas renting households under similar conditions only had \$40,800, showing almost 23 times difference.

Housing appraisal is crucial for the housing market. An accurate appraisal leads to rational negotiation and decision making and thus helps preventing home buyers from buying over-valued homes. Housing appraisal is also highly relevant to financial stability as most banks require a specific house valuation process to decide a healthy mortgage amount. In practice, however, it is difficult for home buyers to access information during house hunting and price negotiation stage because hiring a professional property valuer is expensive, time consuming and inconvenient. These difficulties pose a strong need for a timely, accurate, automatic, and affordable housing appraisal system.

Despite that housing appraisal is important in both macro and micro economics, "housing price remains as much art as science" [18]. The understanding of housing price is still very limited. Existing studies focus on housing attributes (e.g., size of land, number of bedrooms) but largely ignore the relationship between a house and its surroundings. Traditional econometric models have revealed a strong spatial correlation to housing price, which can be explained with two theories [20]: (1) the spillover effect between regions–when physical and human capital, or technological improvement concentrates in one region, it will naturally have a positive impact on its neighbouring regions [9]; (2) unobserved or latent geolocational factors.

Recent availability and appreciation of social, economic and geographic data have enabled researchers to trace the spillover effect on human capital or technological breakthroughs, and to discover unobserved or latent factors. Housing appraisal becomes more viable thanks to granulated geo-spatial rich information available through multiple online resources, such as satellite, street views and housing images, and map data. With the support of real world data, we can empirically investigate how people make decisions. These socioeconomic data sets are multi-source, heterogeneous, and high-dimensional.

Most of the recent development focuses on finding new spatial factors by leveraging the new available online data and machine learning methods to reveal the unexplained elements [15,21,24]. These new findings have shown that housing price is correlated with safer environment [6,8], intangible assets from its neighbourhood, and associations with neighbouring houses [18], design [25], culture [13], Point of Interests measure [11], etc. The new development is mostly from the perspective of neighbourhood characteristics, and is normally within the range of 1 to 2 km from the house. However, only exploring the near neighbourhood has a few limitations. First, certain living functions can't be fulfilled in the near neighbourhood and these functions are not captured in the previous housing price models. For example, shopping malls, hospitals, universities are

---

[2] https://www.abs.gov.au/statistics/people/housing.

strategically located to service at regional level, or national level, not at suburb level. But these functions do influence the demographic distribution in the nearby suburbs and hence influence the housing value. Second, these regional services are economically highly concentrated clusters. For example, a shopping mall can contain 500 shops. A university campus can service 20,000 students. Therefore, economic value and service activities are highly concentrated in these nodes, and would influence the price of houses that are beyond their immediate neighbourhoods.

How can we expand the investigation and identify key factors beyond the immediate neighbourhood?

Few studies [1,16] have shown that enlarging the neighbourhood area can improve housing price prediction. However, these studies were based on satellite image without further investigation of influential visual features, or area beyond the neighbourhood. Therefore, we can't identify whether the improvement is due to merely enlarging the neighbourhood area, or due to the inclusion of new features in the calculation.

To expand the investigation beyond neighbourhood, we can either merely increase the neighbourhood area, or identify and add new key features at the metropolitan level. Increasing the neighbourhood area is not an ideal approach, as it can increase computation significantly without providing additional insights. Therefore, we take the second approach as finding the new key features. It is more challenging, but with the reward of less computation and potentially bringing implication values to home buyers, investors and urban planners.

Our approach extends the relational closeness by investigating the economic proximity. This establishes an economic closeness between the household and the place–economic cluster. This paper aims to study the intangible value of a house beyond neighbourhood value by evaluating the relationship between house and existing regional economic clusters. Specifically, we identify economic clusters by some significant categories, such as CBD, shopping malls, universities. By consolidating with other influencing factors, we build a housing appraisal framework including *Housing* features, *Neighbourhood* characteristics, regional *Economic* clusters and *Demographic* characteristics, called the *HNED* model. This approach may potentially help decision-making for home buyers, property investors and urban planner. It may also indicate solutions for affordable living without compromising the essential needs.

The rest of this paper is structured as follows. Section 2 explains the conceptual framework and main factors in detail. Section 3 explains the experimental settings. Section 4 discusses the results. Section 5 deals with discussion and implications, followed by related work in Sect. 6. Finally, concluding remarks are offered in Sect. 7.

2   Conceptual Framework

In this section, we introduce our housing price estimation model. The target variable of the model is the price of the house. The feature variables are grouped into four feature vectors, corresponding to four different types of attributes, which together influence the price of the house. The types of attributes are: housing attributes, neighbourhood characteristics, regional economic clusters, and demographic characteristics (HNED model).

2.1   Feature Vector 1: Housing Attributes

There are basic attributes about the house itself that people can easily acquire information through advertisement or inspection. These attributes are primary functions that fulfil people's needs of dwelling. The first feature vector corresponds to the influence of these housing attributes. This follows the traditional hedonic model for housing appraisal. There are 18 property attributes selected (corresponding to an 18 dimensional feature vector), which are: area size (total square metres), property type (unit, house, townhouse), number of bedrooms, number of bathrooms, parking, separate study, separate dinning, separate family room, rumpus room, fireplace, walk-in-wardrobe, air condition, balcony, en suite, garage, lockup garage, polished timber floor, barbeque.

2.2   Feature Vectors 2 and 3: The Housing Location

Feature vectors 2 and 3 both relate to the location of the house. In the context of this paper, we interpret location as the accessibility to certain local amenities from the house, as well as the accessibility to important economic clusters from the house, such as the CBD, large shopping centres, universities, etc. Our model captures the quality, quantity and accessibility of the amenities.

There exists rich literature with the investigation of relationship between housing value and its neighbourhood [5–7,10,11]. Different from the current research, we emphasise the value of location rather than neighbourhood. Neighbourhood is the direct geographical and social influence around the individual property. The existing literature neglects the investigation of the relationship between individual property and the regional economic clusters within the metropolitan area. Urban sociologist Burgess [4] emphasised that the urban growth radially expanses from its CBD and physically attractive neighbourhoods. This provides us the theoretical guidance to investigate how location and social networks influence housing prices.

Feature Vector 2: POI-based Neighbourhood Characteristics. The second feature vector focuses on the small POI (Points of Interest) within walking distance (one kilometre), such as shops, restaurants, schools, parks, public transport stops, etc. We consider the location based social network (LBSN) as a cluster of important dots in a map. This location-based feature does not include larger POIs such as shopping centres and universities, which are covered by Feature Vector 3 described in the next section.

Feature vector 2 is a one-dimensional vector. For this feature vector, we assume that all small POI 1 km exerts influence on the price of the house, the quantity of which is one divided by the physical distance between the POI and the house. The total influence on the price of the house by all of the POIs is equal to the sum of the influence by each individual POI. Thus, if we have three POIs at 0.2 km, 0.3 km 0.5 km from the house, the value of Feature 2 for this house is:

$$\frac{1}{0.2} + \frac{1}{0.3} + \frac{1}{0.5} \tag{1}$$

The dataset we use for feature vector 2 was collected from Open Street Maps, which included 13 categories of POIs.

Feature Vector 3: Regional-Level Economic Clusters. Regional-level economic clusters bring new dimension and features into the model. This captures economic activities not happen at neighbourhood level. As discussed in the introduction, capturing all the features in the regional or metropolitan level is time consuming and not realistic. The challenge is to find the most important link between a location and the house. Here we assume the value exchange is the important link. Therefore, we try to capture the economic clusters that provide huge economic values. Our model simplify the Regional Economic Clusters as universities, regional shopping centres, CBD [27] as a preliminary exploration. Features are not limited to these services. Feature Vector 3 is defined in a way with some similarities to Feature Vector 2. Feature Vector 3 is multi-dimensional. The influence of each of the super clusters is weighted by some measure of the size of the cluster (e.g. for shopping centres, the number of shops; for universities, the annual revenue), and attenuated by the inverse of the physical distance between the cluster and the house. Unlike with Feature Vector 2, the economic cluster does not have to be 1 km of the house to exert influence on the housing price.

2.3   Feature Vector 4: Socio-demographic Attributes

Feature Vector 4 involves the socio-demographic attributes of the suburb to which the house belongs, which characterise the social community that is physically closest to the house.

Previous urban economic research has explored the relationships between housing attributes and demographic characteristics of population and found that socio-demographic profiles determines the demand segregation and forms different trends city-wide [27]. We include social-demographic profiles into our modelling for a few reasons. First, we consider social-demographic characteristics can create long-term effect to shape the local economy and community. Housing is not easily transferrable as investment in stock market because of its physical moving difficulty and potential extra capital gain tax for short-term penalty. Residents would stay in the same suburb for a long period and co-create the taste, economy and culture of its local community. The suburb would grow with its residents. Second, human capital can generate externalities and have spill over effect in the neighbourhood regions [9]. Related businesses are more likely to be adjacent and form cluster effect.

Feature vector 4 includes this aspect by adding into the model the relationship between property price and its social-demographic profile. The components of this vector are carefully selected from four sections of the 2016 Australian census data. This include features about income, people and population, education and employment, family and community. We use these features to understand people and their life in each suburb.

Selected features at suburb-level are shown in Fig. 1 with example of one suburb of Melbourne called Brunswick.

Code	Label		
206011105	Brunswick	Median Age - Persons (years)	32.8
		Working Age Population (aged 15-64 years) (%)	79.2
		Persons - Total (no.)	27,435
		Population density (persons/km2)	5,335.0
		Speaks a Language Other Than English at Home (%)	27.9
		Median employee income ($)	50,573
		Mean employee income ($)	59,254
		Median investment income ($)	200
		Mean investment income ($)	4,926
		Total income earners (excl. Government pensions and allowances) (no.)	16,555
		Total income earners (excl. Government pensions and allowances) - median age (years)	34
		Median equivalised total household income (weekly) ($)	1,156
		Completed Year 12 or equivalent (%)	74.8
		Bachelor Degree (%)	29.4
		Unemployment rate (%)	5.9
		Average household size (no. of persons)	2
		Households with mortgage repayments greater than or equal to 30% of household income (%)	4.2
		Households with rent payments greater than or equal to 30% of household income (%)	16.6
		Homeless rate per 10,000 persons (rate)	67.7
		Median commuting distance to place of work (kms)	8.5

Fig. 1. Components of feature vector 4 and sample dataset

3 Experimental Settings

3.1 Data Description

We use metropolitan Melbourne, Australia as our experiment city. The Melbourne housing price data was provided by the Australian Urban Research Infrastructure Network (AURIN). We mainly focus on the sold housing price data in 2018, which includes 161, 179 recorded sold properties. We removed properties whose geographical locations were missing, as well as those whose prices were less than 10, 000. The remaining dataset contains 158, 588 properties.

In order to investigate how location relates to housing price, we collect POIs and public transport stops from OpenStreetMap. We also collect primary school and high school rankings based on their standardised exam results. From TripAdvisor, we collect information of local restaurants and their ranking.

We also collect data to understand regional economic clusters, which include 42 shopping centres and total shop numbers in each centre, 9 universities located in Melbourne and total revenue of each university in 2019.

For the socio-demographic data, we use Australian Bureau of Statistics census data 2016.

3.2 Algorithms

We cast the problem of housing price appraisal in two forms. The first form is estimating the dollar value of the housing price (so the target variable is a continuous variable). The second form is estimating the price range of the house (so the target variable is a categorical variable, representing the estimated price range).

In the dollar value estimation, we use Linear Regression as the baseline algorithm, as this is what is commonly used for such problems in Classical Economics. The performance of three algorithms (Support Vector Machine, Multilayer Perceptron, XGBoost) were compared against the performance of the baseline.

In the price range estimation, we use Logistic Regression as the baseline (classical method), and the performance of four algorithms (Support Vector Machine, Multilayer Perceptron, k-Nearest Neighbour, XGBoost) were compared against the performance of the baseline.

Performance evaluation of the dollar value estimation uses the standard metrics of Mean Absolute Error (MAE), Root-Mean-Squared Error (RMSE), and the Coefficient of Determination (R^2). Performance evaluation of the price range estimation uses the standard metrics for a classification problem of Accuracy, Precision, Recall, and F1.

4   Results and Analysis

4.1   Overall Performance

Table 1 shows the performance of the different algorithms in the task of dollar value estimation. The performance of XGBoost considerably better than the other algorithms, achieving an R^2 value of 0.8779, compared to the baseline performance of 0.6422.

Table 2 shows the performance of the different algorithms in the task of price range estimation. The performance of XGBoost again is considerably better than the other algorithms.

Table 1. Housing price estimation

Model	MAE	RMSE	R^2
Linear regression	0.2450	0.3525	0.6422
Support vector machine	0.4119	0.7620	−0.6723
Multilayer perceptron	0.1994	0.2904	0.7572
XGBoost	**0.1428**	**0.2059**	**0.8779**

Table 2. Housing price range estimation (Classification)

Model	Accuracy	Precision	Recall	F1
Logistic regression	0.6219	0.5432	0.4452	0.4662
Support vector machine	0.5172	0.5162	0.3533	0.3697
k-Nearest neighbour	0.8044	0.7639	0.7631	0.7633
Multilayer perceptron	0.7219	0.6778	0.6145	0.6328
XGBoost	**0.8601**	**0.8590**	**0.8226**	**0.8395**

4.2   The Importance of the Regional Cluster Variable

In this section, we discuss a very important finding that regional cluster variables played a significant role in prediction [14,29]. Table 3 used different combination of subsets of feature vectors. Regional economic cluster as feature vector 3 reached 0.6290 in R^2 individually, and also consistently reached the highest performance when combined with other feature vectors. It is noticeable that feature vectors 1 and 3 combined could reach 0.1552 in MAE, 0.8585 in R^2 which means housing attributes with regional economic cluster variables can give a good prediction.

Table 3. Housing price estimation performance using a subset of the feature vectors

Model	Feature vectors used	MAE	RMSE	R^2
XGBoost	1	0.3614	0.4778	0.3427
XGBoost	2	0.3412	0.4578	0.3965
XGBoost	3	**0.2582**	**0.3589**	**0.6290**
XGBoost	4	0.3214	0.4399	0.4427
XGBoost	1, 2	0.2568	0.3482	0.6508
XGBoost	1, 3	**0.1552**	**0.2217**	**0.8585**
XGBoost	1, 4	0.1875	0.2728	0.7856
XGBoost	2, 3	0.2492	0.3466	0.6540
XGBoost	2, 4	0.2664	0.3680	0.6100
XGBoost	3, 4	0.2507	0.3486	0.6500
XGBoost	1, 2, 3	0.1530	0.2180	0.8632
XGBoost	1, 2, 4	0.1699	0.2447	0.8275
XGBoost	1, 3, 4	**0.1446**	**0.2086**	**0.8747**
XGBoost	2, 3, 4	0.2404	0.3345	0.6778
XGBoost	1, 2, 3, 4	**0.1428**	**0.2059**	**0.8779**

5 Discussions and Implications

Our results show how housing price is related to housing attributes, location and socioeconomic characteristics. Each element contributes different implications for different social agents, such as home buyers, investors, local and regional councils, urban planners.

5.1 Implications for Home Buyers

Generally, home buyers consider both current living functions and investment value of a property. Firstly, housing attributes are the primary focus to meet the daily needs of dwelling. Extra bedroom or bathroom can drive the property value up with better functionality. Secondly, people value more for being in a highly ranked school zone. Walking distance to schools, supermarkets, public transport are preferable by most home buyers. However, the power of strong connection to regional economic clusters may be neglected in the decision making process. To capture a long-term investment return, home buyers need to identify a location with growing highly educated population, with a strong connection to regional economic clusters.

5.2 Implications for Councils and Urban Planning

Based on our results, income and education are strong indicators to drive the housing value, especially investment income is more relevant. High investment income indicates people with multiple source of income or they are business owners. Firstly, councils can attract these group of people by stimulating business district, business park development, or building strong industrial clusters. Secondly, councils can attract young highly educated or highly skilled people to settle down by providing high quality infrastructures, such as high quality public schools, welcoming new campus for highly ranked private schools, or planning shopping centres and sports facilities. Thirdly, councils can make strategic long-term planning of fostering regional economic clusters in a prominent industry, such as forming an IT cluster, education cluster, medical cluster or warehouse cluster, etc. By forming a super cluster, local areas can concentrate human capital in one expertise direction and achieve high economic growth rate.

5.3 Implications for Real-Estate Investors and Developers

Both investors and developers need to identify high demand of housing. Investors focus on both future return and sustainable rental demand. Good location is essential for both rental income and long-term return. Rich POI in the neighbourhood and close distance to regional economic cluster would guarantee a good location. The growth of high educated, skilled population in one area will contribute for future demand of such location and hence drive the future property return. Developers also need to consider maximise the housing needs for potential buyers, extra bedroom and bathroom can significantly increase house value.

6   Related Work

In this section, we discuss the development of housing appraisal methods in both traditional housing market research and computer science research fields. Both fields involved recent development of methods in dealing with the spatial data. Spatial autocorrelation and spatial heterogeneity recognised in both fields are the two main challenges in models involving the spatial data.

In the computer science field, more focus is about how to incorporate newly available data into the house price prediction model. Most of these new data helps to expand the understanding the geographical characteristics towards housing valuation. The nature of these newly available data is often beyond the angle of traditional economic variables, therefore, new methods also introduced into the real estate valuation field. For example, satellite image [1], street-view image, house image from real estate websites [22,25,30] are used for house price prediction [6,16,18]. Other paper investigated how the satellite image and street view can help to understand the neighbourhood [23], demography [12] and commercial activities. These results are also highly relevant to housing appraisal, though they didn't investigate this question directly.

Besides image data, other types of data are used for housing appraisal beyond the traditional scope of economic modeling. Open Street Map data was used for collecting Point of Interests in the neighbourhood [6,11]. Mobile phone data was used to empirically study the human activity and urban vitality [7]. Taxicab trajectory data was mined to estimate neighbourhood popularity, in order to understand the geographic factors for housing appraisal [11]. Google search index was used for housing prediction model to understand how people's attention of real estate would influence the future housing price [28]. Text data from real estate related news was studied to learn how sentiment is related to housing price [17,26].

These data applications are innovative and inspiring for improving housing price modeling. However, most of the newly applied spatial data are focused on the discovery of neighbourhood characteristics, such as crime perception [3], walk-ability [5], cultural influence [13]. These applications neglect the understanding beyond the neighbourhood. And this is the main focus of our work.

7   Conclusions and Future Work

We have studied how factors beyond neighbourhood impact housing values. Specifically, we established regional economic clusters as the significant source of impact beyond neighbourhood. We presented our housing price appraisal model that combined housing attributes, neighbourhood characteristics, and demographic factors. Our model using the XGBoost algorithm has reached 0.88 in R^2, showing the significant impact of regional economic clusters.

Our work enlightens two related research questions worthy of future investigation. (1) Building a customised recommendation system for home buyers. This system aims to tailor and optimise personal needs of affordable living, and provide smart suggestions for trade-offs between different needs and opening up opportunities for locations. (2) Methods to systematically identify regional economic clusters and appropriately weight these clusters in our model. With better understanding of this behavioural mechanism, we could improve our community, facilitate sustainable gentrification, and lead to location diffusion, population growth, and new regional economic clusters emerging.

Acknowledgement. This work has been partly funded by the European Union's Horizon 2020 research and innovation program under the Marie Sklodowska-Curie grant agreement No. 824019 and the DAAD-PPP Australia project "Big Data Security".

References

1. Bency, A.J., Rallapalli, S., Ganti, R.K., Srivatsa, M., Manjunath, B.: Beyond spatial auto-regressive models: Predicting housing prices with satellite imagery. In: 2017 IEEE Winter Conference on Applications of Computer Vision (WACV), pp. 320–329. IEEE (2017)
2. Board, E.S.R.: Vulnerabilities in the EU residential real estate sector (2016)
3. Buonanno, P., Montolio, D., Raya-Vílchez, J.M.: Housing prices and crime perception. Empirical Econ. **45**(1), 305–321 (2013)
4. Burgess, E.W.: The growth of the city: an introduction to a research project. In: Urban Ecology, pp. 71–78. Springer (2008). https://doi.org/10.1007/978-0-387-73412-5_5
5. Cortright, J.: Walking the walk: how walkability raises home values in US cities. CEOs for Cities (2009)
6. De Nadai, M., Lepri, B.: The economic value of neighborhoods: predicting real estate prices from the urban environment. In: 2018 IEEE 5th International Conference on Data Science and Advanced Analytics (DSAA), pp. 323–330. IEEE (2018)
7. De Nadai, M., Staiano, J., Larcher, R., Sebe, N., Quercia, D., Lepri, B.: The death and life of great italian cities: a mobile phone data perspective. In: Proceedings of the 25th International Conference on World Wide Web, pp. 413–423 (2016)
8. De Nadai, M., et al.: Are safer looking neighborhoods more lively? A multimodal investigation into urban life. In: Proceedings of the 24th ACM International Conference on Multimedia, pp. 1127–1135 (2016)
9. Ertur, C., Koch, W., et al.: Convergence, human capital and international spillovers. Laboratoire d'Economie et de Gestion Working Paper (2006)
10. Fu, Y., et al.: Sparse real estate ranking with online user reviews and offline moving behaviors. In: 2014 IEEE International Conference on Data Mining, pp. 120–129. IEEE (2014)
11. Fu, Y., Xiong, H., Ge, Y., Yao, Z., Zheng, Y., Zhou, Z.H.: Exploiting geographic dependencies for real estate appraisal: a mutual perspective of ranking and clustering. In: Proceedings of the 20th ACM SIGKDD International Conference on Knowledge Discovery and Data Mining, pp. 1047–1056 (2014)
12. Gebru, T., et al.: Using deep learning and google street view to estimate the demographic makeup of neighborhoods across the united states. Proc. Natl. Acad. Sci. **114**(50), 13108–13113 (2017)
13. Hristova, D., Aiello, L.M., Quercia, D.: The new urban success: how culture pays. Front. Phys. **6**, 27 (2018)

14. Huang, J., Peng, M., Wang, H., Cao, J., Gao, W., Zhang, X.: A probabilistic method for emerging topic tracking in microblog stream. World Wide Web 20(2), 325–350 (2016). https://doi.org/10.1007/s11280-016-0390-4
15. Jiang, H., Zhou, R., Zhang, L., Wang, H., Zhang, Y.: Sentence level topic models for associated topics extraction. World Wide Web 22(6), 2545–2560 (2018). https://doi.org/10.1007/s11280-018-0639-1
16. Kostic, Z., Jevremovic, A.: What image features boost housing market predictions? IEEE Trans. Multimed. 22(7), 1904–1916 (2020)
17. Kou, J., Fu, X., Du, J., Wang, H., Zhang, G.Z.: Understanding housing market behaviour from a microscopic perspective. In: 2018 27th International Conference on Computer Communication and Networks (ICCCN), pp. 1–9. IEEE (2018)
18. Law, S., Paige, B., Russell, C.: Take a look around: using street view and satellite images to estimate house prices. ACM Trans. Intell. Syst. Technol. (TIST) 10(5), 1–19 (2019)
19. Leamer, E.E.: Housing is the business cycle. Technical Report National Bureau of Economic Research (2007)
20. LeSage, J.P.: An introduction to spatial econometrics. Revue d'économie industrielle 123, 19–44 (2008)
21. Li, H., Wang, Y., Wang, H., Zhou, B.: Multi-window based ensemble learning for classification of imbalanced streaming data. World Wide Web 20(6), 1507–1525 (2017). https://doi.org/10.1007/s11280-017-0449-x
22. Liu, X., Xu, Q., Yang, J., Thalman, J., Yan, S., Luo, J.: Learning multi-instance deep ranking and regression network for visual house appraisal. IEEE Trans. Knowl. Data Eng. 30(8), 1496–1506 (2018)
23. Naik, N., Kominers, S.D., Raskar, R., Glaeser, E.L., Hidalgo, C.A.: Computer vision uncovers predictors of physical urban change. Proc. Natl. Acad. Sci. 114(29), 7571–7576 (2017)
24. Peng, M., Zeng, G., Sun, Z., Huang, J., Wang, H., Tian, G.: Personalized app recommendation based on app permissions. World Wide Web 21(1), 89–104 (2017). https://doi.org/10.1007/s11280-017-0456-y
25. Poursaeed, O., Matera, T., Belongie, S.: Vision-based real estate price estimation. Mach. Vision Appl. 29(4), 667–676 (2018). https://doi.org/10.1007/s00138-018-0922-2
26. Soo, C.K.: Quantifying sentiment with news media across local housing markets. Rev. Financ. Stud. 31(10), 3689–3719 (2018)
27. Thériault, M., Des Rosiers, F., Villeneuve, P., Kestens, Y.: Modelling interactions of location with specific value of housing attributes. Property Manage. 21, 25–62 (2003)
28. Wu, L., Brynjolfsson, E.: The future of prediction: how google searches foreshadow housing prices and sales. In: Economic Analysis of the Digital Economy, pp. 89–118. University of Chicago Press (2015)
29. Yin, J., Tang, M.J., Cao, J., Wang, H., You, M., Lin, Y.: Adaptive online learning for vulnerability exploitation time prediction. In: Huang, Z., Beek, W., Wang, H., Zhou, R., Zhang, Y. (eds.) WISE 2020. LNCS, vol. 12343, pp. 252–266. Springer, Cham (2020). https://doi.org/10.1007/978-3-030-62008-0_18
30. You, Q., Pang, R., Cao, L., Luo, J.: Image-based appraisal of real estate properties. IEEE Trans. Multimed. 19(12), 2751–2759 (2017)

Modeling Daily Crime Events Prediction Using Seq2Seq Architecture

Jawaher Alghamdi$^{(\boxtimes)}$ and Zi Huang

School of Information Technology and Electrical Engineering,
The University of Queensland, Brisbane, Australia
j.alghamdi@uqconnect.edu.au, huang@itee.uq.edu.au

Abstract. Early prediction of the crime occurrence reduces its impact. Several studies have been conducted to forecast crimes. However, these studies are not highly accurate, particularly in short-term forecasting such as over one week. To respond to this, we examine sequence to sequence (Seq2Seq) based encoder-decoder LSTM model using two real-world crime datasets of Brisbane and Chicago, extracted from the open data portal, to make one week ahead of total daily crime forecasting. We have built an ARIMA statistical model and three machine learning-based regression models that differ in their architecture, namely, simple RNN, LSTM, and Conv1D with a novel approach of walk-forward validation. Using a grid search strategy, the hyperparameters of the models are optimized. The obtained results demonstrate that the proposed Seq2Seq model is highly effective, if not superior, compared to its counterparts and other algorithms. This proposed model achieves state-of-the-art results with a relatively Root Mean Squared Error (RMSE) of 0.43 and 0.86 on both datasets, respectively.

Keywords: Crime events forecasting · Sequence to sequence

1 Introduction

Time series data can be defined as a combination of observations of the variable's values at different regular time intervals. Such data must exhibit a sequential manner, and the order must exist between the data points. Time series forecasting has recently attracted researchers' attention and has been applied extensively in many domains, particularly criminology. Furthermore, time series analysis has become an essential approach to model the criminal temporal behaviour where meaningful insights could be extracted through analysing sequences of temporally observed data. In recent years, crime prediction has received great attention because of its benefit since it is crucial to provide guidance for people's lives and help for crime incidents control. While traditional univariate forecasting techniques such as ARIMA are extensively used in the time series forecasting approach due to the simplicity, they might fail to handle a tremendous amount of data, resulting in unreliable prediction. On the other hand, utilising machine

© Springer Nature Switzerland AG 2021
M. Qiao et al. (Eds.): ADC 2021, LNCS 12610, pp. 192–203, 2021.
https://doi.org/10.1007/978-3-030-69377-0_16

learning in the context of big data would typically yield better results. With widespread criminal events, there is an increasing emphasis for predictive models that can accurately and effectively handle time-series crime data to forecast future criminal incidents. Time series values have an inherent sequential format, and so traditional neural networks cannot be used to map sequences to sequences as they cannot maintain information about previous events in a sequence of events. Thus, to handle time-series data, Hihi [8] proposed recurrent neural networks (RNNs) where it is easier for the recurrent neural network to map sequences to sequences [11,12]. Driven by the recurrent neural network (RNN) architectures with long short-term memory (LSTM) units, neural networks have achieved state-of-the-art results in several prediction tasks. Despite RNNs remarkable success, it experiences problems related to gradient vanishing and exploding, making it difficult to handle long sequences – meaning they break down during training and fail to model the problem well [13]. Later, long short-term memory networks (LSTMs) were introduced and explicitly designed to overcome this limitation [10]. However, it is still hard for those algorithms mentioned above to support different length sequences. Cho et al. [6] developed sequence to sequence model to respond to this problem. This paper aims to conduct short-term forecasting of crime count at a daily level in the Brisbane and Chicago cities utilising historical data. Forecasting the total number of crimes at a day level would help police and law enforcement agencies to prepare and distribute the resources precisely and efficiently. We utilised real-world crime datasets for two cities to implement our models. It is noteworthy to mention that Brisbane data has been collected using the provided API in qld.police.au website. Because sequence-to-sequence based encoder-decoder model has not been applied within the criminology domain before, and because of their remarkable success in solving similar problems in several domains as well as its exceptional capability in multi-steps forecasting, it raises the question of whether such an approach can be used to predict the sequence of the total daily crime events for the subsequent week. Expressed in a more concise manner, we attempt to answer the question:

Can we use Seq2Seq learning to model the problem of forecasting the future week of total daily crime? If so, how does the Seq2Seq model compare to the traditional algorithms. Therefore, our contribution is summarised as follows:

- We propose Seq2Seq based encoder-decoder LSTM model to predict the subsequent week of total daily crime in Brisbane and Chicago cities.
- We found that the Seq2Seq model is significantly effective for such a prediction. It achieves state-of-the-art performance with a Root Mean Squared Error (RMSE) of 0.43 and 0.86 for Brisbane and Chicago data, respectively.

The rest of this paper is organised as follows. Section 2 summarises the literature related to the prediction of crime. In Sect. 3, we describe our research methodology and the proposed Seq2Seq model. Extensive results on the performance of the predictive models are presented in Sect. 4. This section describes all other predictive models built in this work. Finally, Sect. 5 concludes the paper.

2   Related Work

Time series of crime data modelling is inherently challenging due to the fact that crime is a randomly occurring pattern. A large body of work currently exists designed to predict future crime. Because increasing the time granularity is useful and helpful, previous related work concentrated on long time intervals such as a month and years to predict aggregated crime trends [9,15]. Classical statistical techniques have been applied to many crime prediction tasks. A study by [4] applied auto-regressive model approach for forecasting the number of crimes that will happen in rolling time horizons in urban areas. Another study by [5] uses the auto-regressive moving average (ARIMA) model to forecast crime at a weekly level for a city in China. In [14], the authors examined the fuzzy time series technique for crime prediction. In their study, they utilised seventeen years of historical crime data of Delhi city. However, the linear property of the statistical methods used for time series forecasting such as Box-Jenkins approaches make it difficult to approximate any nonlinear functions and to address this, neural networks were explored. Machine learning approaches have shown remarkable success and have been applied to future crime prediction. The work of Yu et al. [17] aimed at predicting burglary crime type hotspots at a monthly level utilising a shallow feed-forward network. Similarly, a study by Bogomolov et al. [2], the authors use a simple shallow feed-forward neural network to predict the crime cascade in London areas on a monthly time scale. In this work, we aim to forecast the next week of total daily crime since this would help the police to be much more accurate with their efforts. We exploit the power of machine learning algorithms, and we proposed Seq2Seq model, and we studied its performance in predicting the next week of total daily crime in the future.

3   Methodology

In this section, we briefly discussed how we collected and prepared our datasets. Then, we presented how we constructed the proposed predictive model to achieve the objective of this research.

3.1   Data Collection

In order to build the predictive models, Chicago historical crime data were collected from the Chicago open data portal, which involves all reported criminal incidents from 2001 to July 2020. The dataset consists of a total of more than seven million records and twenty-two features, including the exact locations (longitude and latitude), the type of crimes committed (a total of 34 types of crime events), community areas, districts, wards where the incidents occurred and other features. Brisbane crime records, on the other hand, are available from the qld.police.au website – which were collected through the API provided in the website. The collected data covered the periods between November 2015 and July 2020 and consisted of 389354 records. The geographical location, date, and crime type (a total of 21 types of crime events) are provided for each crime event. For both datasets, we consider the data from 2015 to 2020 for multiple reasons. Brisbane data spans from 2015 to 2020, and we decided to consider the period from 2015 to 2020 for Chicago. Another reason is that crimes shift over time, and as time goes on, old records will become less relevant [15].

3.2   Data Pre-processing

Pre-processing is a significant task in data science solution, as we need to create a cleaned data to train the machine using algorithms. Considering the fact that better quality data results in better prediction performance, a fundamental step in the data science process is the preprocessing step. To best frame our time series task, we performed a preprocessing step to create a new feature wherein we aggregate the records by the date attribute in order to obtain a sequential data of daily crime data that is appropriate for the forecasting task. Consequently, the resulting data, which encompasses single observations (total-crimes) recorded sequentially, describes the daily number of crimes over five years, which implies that our task is a univariate time series forecasting. Because the sequence order of the time series must be maintained, as such, this data is partitioned sequentially in a format of standard weeks, for all models unless explicitly stated otherwise, into training and testing data. Similar to what [3] did, we select the first four years for training the predictive models and the last year for testing models; more precisely, for Chicago dataset, we have a total of 289 weeks from Jan. 2015 to July 2020. The data is divided into standard weeks (the week starts in Sun. and ends in Sat.) for both training (a total of 242 weeks) and testing sets (a total of 47 weeks). On the other hand, for Brisbane crime data, the data spans from 2015 to 2020 as well; we follow the same procedure, dividing the data into standard weeks. The data starts in late Nov. 2015, and the first Sun. in the data is 8 Nov. 2015, and so training data starts from eight and finish in 24th, Aug. 2019 which is Saturday and that gives a total of 198 weeks. For the testing set, it starts from Sunday 25 Aug. 2019 to Saturday 18 Jul. 2020. To train machine learning models, a sliding window process to convert the time-series sequence into input-output pairs (supervised learning) is to be implemented so as to prepare the data for the modelling. This can be done by iterating and dividing the data in question according to the pre-defined time-steps. At each time step, seven observations will be taken as input, and the subsequent seven observations will be considered an output sequence.

3.3   Proposed Seq2seq Model

Problem Description. Time series forecasting has been extensively applied to gain useful insights and knowledge of the future. Univariate time series forecasting, as the name suggests, is defined as a series with a one-time dependent attribute where the previous values of the time series are used to predict the future values. We firstly will apply classical time series approaches such as ARIMA. Then, we will move to machine learning approaches, in which the task is treated as a sequence to a scalar problem, such as simple recurrent neural network (RNN), long short-term memory (LSTM), 1D convolutional neural network (Conv1D) as well as our proposed Seq2Seq based encoder-decoder model. The models take input data in a weekly format and output the subsequent week of seven days of total daily crimes as forecasted values.

Model Architecture. The proposed Seq2Seq-based crime forecasting model follows the original Seq2Seq architecture comprising of encoder and decoder blocks. Seq2Seq modelling has been applied with recurrent neural network-based encoder-decoder architectures [1,16]. As the name suggests, the input to Seq2Seq model and the output are both sequence of items. The model architecture encompasses two main blocks, namely, encoder and decoder. The encoder can be any of recurrent units; however, a common choice for RNNs based encoder-decoder models is LSTM [10]. Therefore, to figure out the intermediate encoder states z and decoder states h we utilise long short-term memory (LSTM) where a sequence of input $x = (x_1, ..., x_n)$ of n items – single item at a time, is processed and a sequence of state representations $z = (z_1, ..., z_n)$ is generated. Moreover, this encoded z vector is often referred to as a context vector – the final hidden state generated from the encoder, which is then taken as an input to the decoder to produce the output sequence $y = (y_1, ..., y_n)$. In this work, both encoder and decoder made of a single LSTM. At each time step, every single recurrent unit takes an input element from a sequence of elements and does some processing, updates its hidden state based on its input and the previous inputs in the sequence and generates the output for that time step. As such, Seq2Seq model aims to teach a model (in this case, several LSTM units) to forecast the future crime count. To apply machine learning on sequence data is challengeable due to the fact that the dimensionality of input and output must be fixed and known in advance [16]. To respond to this problem, two recurrent neural networks-based LSTM will be implemented whereby the first LSTM is used to process the sequence of input in such a way and encode them in one fixed length vector called context vector – last hidden state of the encoder LSTM. The second LSTM is then utilised to decode such given vector to output the sequence [16].

Training and Implementation. The univariate Seq2Seq model does not produce a vector sequence as its output directly. Instead, the model will produce the prediction for every single day at each timestep, not for all the seven days in a single round. The LSTM encoder layer is built with 200 units which output 200 element vector, each of which constitutes one output that captures features from the seven-value input sequence. The main purpose of the encoder is to encode the input sequence – with the shape (7, 1) which implies that the time series is univariate with one feature being considered for the prior one week's data as input, after processing them into a fixed-length vector which is the last hidden state of the encoder. It is noteworthy to mention that we care more about the final output generated at the end of the sequence in the encoder and ignore the output produced at each time step.

The internal representation of the input sequence is repeated several times for each timestep in the output sequence. This can be achieved by using repeat vector layer that acts as an adapter for adjusting the dimension of the data and thus connects both blocks together [3]. The repeat vector layer's data shape is (7, 200), which corresponds to the seven timestamps in the output sequence.

Another LSTM layer with 200 nodes is added, which acts as a decoder to decode the encoded input sequence. The decoder has the same structure of the LSTM used in the encoder; however, in contrast to the encoder, the entire sequence will be generated (i.e., a value for each day of the seven days will be generated) to enable the decoder to have a piece of knowledge about the previous predicted day and accumulate an internal state while outputting the sequence [3]. In other words, each of those nodes will produce a value for each of the seven days in a week. The decoder's output sequence is then passed through a fully connected layer, which in turn interprets each value in the output sequence. Finally, the output layer will generate the prediction for each day of the output sequence, as mentioned previously, single day at a time. In order to interpret and process each time step in the output sequence, the fully connected layer and the output layer were wrapped using TimeDistributed wrapper. We use ReLU as an activation function in the middle layers of the neural network. Contrary to LSTM and Conv1D models, the output of Seq2Seq model is a three-dimensional vector where each output has the shape of $[n_{samples}, n_{timestamps}, n_{features}]$.

The gradient is normalized utilizing Adam optimizer with $\beta_1 = 0.9$, $\beta_2 = 0.999$, $\epsilon = 1e-07$ and $lr = 0.001$ and mean squared error as a loss function were used at the final output layer of the model. The loss formula is as follows:

$$loss = \frac{1}{m} \sum_{i}^{m} |(y_{truth}^i - y_{pre}^i)|^2, \tag{1}$$

where y_{truth}^i is the actual value, and y_{pre}^i is the predicted value.

For the training process, we train for 250 and 100 epochs for Brisbane and Chicago data, respectively. Empirically, we set the sliding window size for the input and the output as seven timesteps (we experiment with different input timesteps as will be discussed next).

4   Results and Discussions

4.1   ARIMA, Simple RNN, LSTM and Conv1D

ARIMA. In the case of a univariate ARIMA model, the underlying assumption is that each week of total daily crimes relies on the prior seven days and prior q seven days estimation errors. Therefore, the grid search strategy was applied with 5-Fold cross-validation using the following parameter space: p, q in range $(0, 9)$, and d in range $(0, 3)$, in order to select the best that yields the best performance on cross-validation experiment. Then, the most suitable candidate models were selected – which are the models for the order (p, d, q) equal to (3, 0, 0) and (4, 0, 0) for Brisbane and Chicago data, respectively. We chose Akaike's

Information Criterion (AIC) as a metric – use Eq. (2), as well as Mean Squared Error. Good models are selected by minimising those metrics.

$$AIC = -2log(L) + 2(p + q + k + 1) \tag{2}$$

Simple RNN. We applied the following settings for both datasets: the RNN model has a single RNN layer with 200 units, a fully connected layer with 100 units and a final dense layer which outputs a vector of size seven, which is equal to the number of steps ahead predictions. An activation function called Exponential Linear Unit 'ELU' is being chosen not only because of its ability to generate accurate crime count forecasting results but also because it helps to converge cost to zero faster [7]. Mean squared error as a loss function and Adam optimizer as an optimization function are used for training.

LSTM. Using the univariate sequence of the total daily crime data, the LSTM model that uses the prior one-week data as the input makes multi-step time series forecasting. The input shape to the network is (7, 1) means that the previous seven values of the series (i.e., one-week data) are used as input, and only one attribute (i.e., total crimes) is considered since our task framed as a univariate problem. The input layer transfers the data to the LSTM layer with 200 nodes at the output, and in those nodes, the ReLU activation function is used. The LSTM layer is followed by a dense layer with 200 nodes at its input, and 100 nodes at the output with the ReLU activation function. The dense layer is then connected to the output layer with 100 nodes at its input and seven nodes at the output. The nodes at the output layer use MSE as a loss function and Adam as an optimizer.

Conv1D. As the name suggests, Conv1D encompasses a conventional hidden layer that works on a one-dimensional sequence. We applied a Conv1D model with 16 and 32 filter size in Brisbane and Chicago data, respectively, and three number windows to be learned where the values in such a window will be multiplied by the filter values. A max-pooling layer is added, followed by a dense layer in order to represent the input features. We utilized a flatten layer in between in order to reduce the feature maps to a single one-dimensional vector. As our task is a univariate time series, the input shape would be reshaped to be the number of the pre-defined time-steps size and the feature size which is, in this case, one variable. The model is then fitted with Adam optimizer and MSE as a loss function. We propose sequence to sequence, Seq2Seq in short, for the task of predicting next week of total daily crimes. We use Simple RNN, LSTM, Conv1D, and a statistical ARIMA method as baselines to our proposed model. We apply two experiments: (1) We experimented using different timesteps for machine learning models to compare the performance and choose the best number of timesteps as an input. (2) To evaluate the benefits of proposed Seq2Seq model, we compare against the baselines mentioned above using two types of evaluation metrics, namely, Mean Absolute Error (MAE) as well as Root Mean Squared

Error (RMSE) – use Eq. (3), respectively. We follow multi-step forecasting with walk-forward validation approach [3], which states that the prior last week of training is used to predict the next week of testing data and the actual records for that week were added to the training dataset in order to be used to forecast the next week of total daily crimes. The obtained results for each algorithm will be discussed in the following subsections.

$$MAE = \frac{1}{n}\sum_i |(y_i - \hat{y}_i)|, RMSE = \sqrt{\frac{1}{n}\sum_i (y_i - \hat{y}_i)^2}, \tag{3}$$

where $\hat{y}$ denotes the predicted value of y and y denotes the actual value. The models exhibiting lower values of those metrics are supposed to be more accurate.

4.2   Prediction Performance Comparison Using Different Timesteps

For both datasets, we experiment using different timesteps (number of prior observations) to test the models' predictive performance for predicting the subsequent week of total daily crimes. Given the essence of autocorrelation in a time series, the number of lagged observations used in neural networks is often a more significant factor than the number of hidden nodes [19].

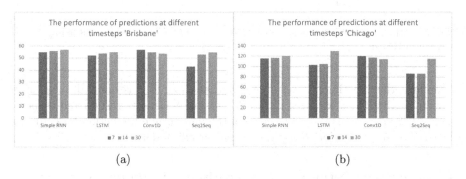

Fig. 1. The performance of predictions at different timesteps using RMSE on (a) *Brisbane* and (b) *Chicago* datasets.

Consequently, as can be seen from Fig. 1, for both time-series datasets, as the number of timesteps decreases so does the error, indicating that the prior observations that are closer in the sequence have a more significant impact than those further away from our target sequence. More specifically, using the prior week to predict the next week yields better results for all models except for the prediction performance of Conv1D where we agree with [20] that the prediction performance of Conv1D shows better performance as the number of timesteps increase. Our proposed Seq2Seq model shows superior results at timestep = 7 compared to other input timesteps, according to RMSE as an evaluation metric.

Table 1. Performance comparison (%) of overall MAE and RMSE across all seven days on the *Brisbane* dataset.

	MAE	RMSE
ARIMA	39.5	51.3
Simple RNN	39	55.3
LSTM	37.8	52.6
Conv1D	41.7	58.9
Seq2Seq	**28.7**	**43.1**

Table 2. Performance comparison (%) of overall MAE and RMSE across all seven days on the *Chicago* dataset.

	MAE	RMSE
ARIMA	78.3	109.7
Simple RNN	73	116.8
LSTM	64.3	102.6
Conv1D	73.1	114.2
Seq2Seq	**53.8**	**86.2**

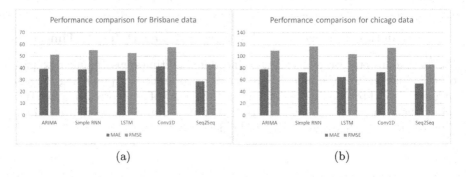

(a)   (b)

Fig. 2. Performance comparison (%) for *Brisbane* and *Chicago* datasets using overall MAE and RMSE.

Therefore, the experimental results prove that using the historical data of seven days to predict the next seven days will give more accurate results and it shows that Seq2Seq model is the best choice of the daily crime prediction task. In this essence, we use the prior week as an input for all models.

4.3   Prediction Performance Comparison with Baselines

The results of the proposed Seq2Seq model and the four baseline models on Brisbane and Chicago datasets are tabulated in Table 1 and Table 2, respectively. As mentioned above, Mean Absolute Error (MAE) and Root Mean Squared Error (RMSE) are used to evaluate different models' performance. The best performance for each evaluation metric is highlighted in bold.

Remarkably, we can see from the results in Fig. 2, that the proposed Seq2Seq model consistently performs better than all four baselines on both datasets across all evaluation metrics. Interestingly, the days that acquired the least RMSE are Wednesday and Thursday in both datasets. This can be seen in Fig. 3. Since Chicago data shows a clear seasonality structure (periodic cycles and recurrent patterns), this, in fact, may have a negative effect on various forecasting methods. Compared to the seasonal ARIMA model, sequence to sequence model trained

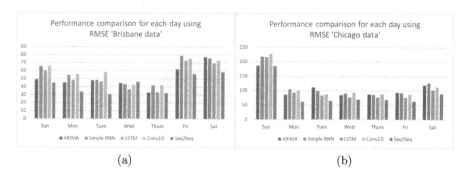

Fig. 3. Performance comparison (%) for each day using RMSE on both datasets.

with both datasets consistently forecast better than ARIMA and other models across all performance measures, as shown in Fig. 2. For Chicago data, other machine learning algorithms built with original data perform worse than the ARIMA model. While some studies argued that deseasonalization preprocessing is unnecessary for machine learning algorithms since they can capture and model seasonality directly, others just disagreed with that [18]. However, the obtained results suggest the importance of further preprocessing for the data in question before feeding it to the models. We believe that although the ARIMA model with deseasonalization preprocessing outperforms other machine learning models, it does not always beat machine learning algorithms if further appropriate data preprocessing is introduced (although we did not verify it experimentally). Therefore, given the results reported above, it is possible that a preprocessing step of deseasonalization can be considered to improve the performance. This indicates that there still a room for improving machine learning models using data that exhibits seasonal structure. In this study, although ARIMA might well be improved by applying further hyperparameter optimization to select the optimal values for the parameters (p, d, q) as well as the LSTM, and Conv1D by adding more crime data that may have led to a lower prediction error, Seq2Seq is the preferred model, at least at this analysis level. With the fitting results, it is believed that Seq2Seq model makes better accurate forecasting. In order to verify this, daily crime count of one week ahead is predicted with all models, and their results are compared with the actual values on both Brisbane and Chicago data – see the outcomes in Fig. 4. As can be seen from the figures, Seq2Seq model estimates very well the real observations and has an exceptional capability to capture the series fluctuations, which implies that Seq2Seq can generalize very well on unseen data. Generally speaking, machine learning models on both datasets outperform ARIMA model for future prediction. To this end, we can draw an answer to our question – via the performance comparison, the proposed Seq2Seq model outperforms representative baselines for one week ahead of total daily crime prediction. The obtained accurate results by our proposed Seq2Seq model are beneficial and crucial for the police in preparing the patrolling in advance and the scarce resources to be deployed efficiently.

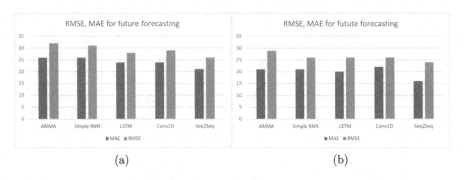

(a)   (b)

Fig. 4. Performance comparison (%) for future forecasting using overall MAE and RMSE on (a) *Brisbane* and (b) *Chicago* datasets.

5   Conclusion

In this paper, we take the prediction of crime count as a univariable time series forecasting problem. We have constructed and optimized several approaches to predict the subsequent week of total daily crimes using two real-world crime datasets of Brisbane and Chicago that span from 2015 to 2020. Among all the predictive models, the proposed Seq2Seq model's performance was found to be far too superior to that of the other models. The study reveals that the proposed model can capture and learn long-term dependencies from time-series data. Furthermore, the experimental results show that the proposed model could reasonably predict the crime in the subsequent week in the future period where it is generated the minimum MAE and RMSE compared to the other techniques. As a future work scope, we will investigate the possibility of adding the attention mechanism to seq2seq architecture in time series forecasting of crimes.

References

1. Bahdanau, D., Cho, K., Bengio, Y.: Neural machine translation by jointly learning to align and translate. arXiv preprint arXiv:1409.0473 (2014)
2. Bogomolov, A., Lepri, B., Staiano, J., Oliver, N., Pianesi, F., Pentland, A.: Once upon a crime: towards crime prediction from demographics and mobile data. In: Proceedings of the 16th International Conference on Multimodal Interaction, pp. 427–434 (2014)
3. Brownlee, J.: Deep learning with time series forecasting. Machine Learning Mastery (2018)
4. Cesario, E., Catlett, C., Talia, D.: Forecasting crimes using autoregressive models. In: 2016 IEEE 14th International Conference on Dependable, Autonomic and Secure Computing, 14th International Conference on Pervasive Intelligence and Computing, 2nd International Conference on Big Data Intelligence and Computing and Cyber Science and Technology Congress (DASC/PiCom/DataCom/CyberSciTech), pp. 795–802. IEEE (2016)

5. Chen, P., Yuan, H., Shu, X.: Forecasting crime using the ARIMA model. In: 2008 Fifth International Conference on Fuzzy Systems and Knowledge Discovery, vol. 5, pp. 627–630. IEEE (2008)
6. Cho, K., et al.: Learning phrase representations using RNN encoder-decoder for statistical machine translation. arXiv preprint arXiv:1406.1078 (2014)
7. Clevert, D.A., Unterthiner, T., Hochreiter, S.: Fast and accurate deep network learning by exponential linear units (ELUs). arXiv preprint arXiv:1511.07289 (2015)
8. El Hihi, S., Bengio, Y.: Hierarchical recurrent neural networks for long-term dependencies. In: Advances in Neural Information Processing Systems, pp. 493–499 (1996)
9. Feng, M., Zheng, J., Han, Y., Ren, J., Liu, Q.: Big data analytics and mining for crime data analysis, visualization and prediction. In: Ren, J., et al. (eds.) BICS 2018. LNCS (LNAI), vol. 10989, pp. 605–614. Springer, Cham (2018). https://doi.org/10.1007/978-3-030-00563-4_59
10. Hochreiter, S., Schmidhuber, J.: Long short-term memory. Neural Comput. 9(8), 1735–1780 (1997)
11. Jozefowicz, R., Zaremba, W., Sutskever, I.: An empirical exploration of recurrent network architectures. In: International Conference on Machine Learning, pp. 2342–2350 (2015)
12. Lipton, Z.C., Berkowitz, J., Elkan, C.: A critical review of recurrent neural networks for sequence learning. arXiv preprint arXiv:1506.00019 (2015)
13. Pascanu, R., Mikolov, T., Bengio, Y.: On the difficulty of training recurrent neural networks. In: International Conference on Machine Learning, pp. 1310–1318 (2013)
14. Shrivastav, A.K., Ekata, D.: Applicability of soft computing technique for crime forecasting: a preliminary investigation. Int. J. Comput. Sci. Eng. Technol. 9(9), 415–421 (2012)
15. Stec, A., Klabjan, D.: Forecasting crime with deep learning. arXiv preprint arXiv:1806.01486 (2018)
16. Sutskever, I., Vinyals, O., Le, Q.V.: Sequence to sequence learning with neural networks. In: Advances in Neural Information Processing Systems, pp. 3104–3112 (2014)
17. Yu, C.H., Ward, M.W., Morabito, M., Ding, W.: Crime forecasting using data mining techniques. In: 2011 IEEE 11th International Conference on Data Mining Workshops, pp. 779–786. IEEE (2011)
18. Zhang, G.P., Qi, M.: Neural network forecasting for seasonal and trend time series. Eur. J. Oper. Res. 160(2), 501–514 (2005)
19. Zhang, G.: Linear and nonlinear time series forecasting with artificial neural networks. Kent State University (1998)
20. Zhao, L., Cheng, B., Chen, J.: A hybrid time series model based on dilated Conv1D and LSTM with applications to PM2. 5 forecasting. Aust. J. Intell. Inf. Process. Syst. 17(2), 49–60 (2019)

Modelling and Factorizing Large-Scale Knowledge Graph (DBPedia) for Fine-Grained Entity Type Inference

A. B. M. Moniruzzaman[(⊠)]

University of Technology Sydney, Sydney, Australia
abm.moniruzzaman@student.uts.edu.au

Abstract. Recent years have witnessed a rapid growth of knowledge graphs (KGs) such as Freebase, DBpedia, or YAGO. These KGs store billions of facts about real-world entities (e.g. people, places, and things) in the form of triples. KGs are playing an increasingly important role in enhancing the intelligence of Web and enterprise search and in supporting information integration and retrieval. Linked Open Data (LOD) cloud interlinks KGs and other data sources using the W3C Resource Description Framework (RDF) and makes accessible on web querying. DBpedia, a large-scale KG extracted from Wikipedia has become one of the central interlinking hubs in the LOD cloud. Despite these impressive advances, there are still major limitations regarding coverage with missing information, such as type, properties, and relations. Defining fine-grained types of entities in KG allows Web search queries with a well-defined result sets. Our aim is to automatically identify entities to be semantically interpretable by having fine-grained types in DBpedia. This paper embeddings entire DBpedia, and applies a new approach based on a tensor model for fine-grained entity type inference. We demonstrate the benefits of our task in the context of fine-grained entity type inference applying on DBpedia, and by producing a large number of resources in different fine-grained entity types for connecting them to DBpedia type classes.

Keywords: Knowledge graph · DBPedia embedding · Type inference · Tensor factorization · Semantic web search

1   Introduction

Knowledge graphs (KGs), i.e., graph-based knowledge-bases, store information about real-world objects (e.g. people, places, and things) in the form of RDF triples (i.e. (subject, predicate, object)). Recent years have witnessed a rapid growth of KGs driven by academic and commercial efforts, such as Yago [26, 49], Freebase [13], DBpedia [10,36], NELL [15], Google's Knowledge Graph, Microsoft's Satori, Probase [3], and Google Knowledge Vault [25]. These KGs have reached an impressive size, for instance, DBPedia a large-scale KG extracted

© Springer Nature Switzerland AG 2021
M. Qiao et al. (Eds.): ADC 2021, LNCS 12610, pp. 204–219, 2021.
https://doi.org/10.1007/978-3-030-69377-0_17

from Wikipedia contains many millions of entities, organized in hundreds to hundred thousands of semantic classes, and billions of relational facts (triples) involving a large variety of predicates (relation types) between entities.

KGs are playing an increasingly important role in enhancing the intelligence of Web and enterprise search and in supporting information integration and retrieval. For Example, Freebase KG powered Google Knowledge Graph that supports Google's web search, or Microsoft's Satori that supports Bing by providing richer data for Entity Pane, Carousel, and Facts Across Segments in the search panel. Additionally, KGs are becoming important resources for different Artificial Intelligence (AI) and Natural Language Processing (NLP) applications, such as Question-Answering [11,22], Query Understanding through Knowledge-Based Conceptualization [12], and Short Sentence Texts Understanding [51,53] and Conceptualization using a probabilistic Knowledge bases. Despite these impressive advances, there are still major limitations regarding coverage and freshness, these KGs are incomplete with missing information, such as type, properties, and relations [18,42,45,48,52,57].

Types in KG are used to express the concept of classes. According to KG idiomatic usage, a KG object "has X, Y, Z types" is equivalent to an object "is a member of the X, Y, Z classes". In the case of Tom Hanks[1] , the KG object for Hanks would have the types *person* and *Actor* to indicate that the object is a member of the Persons and Actors. However, an entity is usually not associated to a limited set of generic types (Person, Location, and Organization) in KGs but rather to a set of more specific (fine-grained) types. Evidence suggests that performance of Web search queries (in case of exploring lists and collections) can be dramatically improved by defining large numbers of these fine-grained entity types in KG. Untyped entities and entities with incomplete set of types are a common problem in Semantic Web KGs [42,45]. For example, one can find by Web search queries the fact that Tom Hanks is a person, an actor, and a person from California, USA. All these types are correct but some may be too general to be interesting (e.g., person, actor), while other set of more specific (fine-grained) types may be interesting but may be identified by web searching (such as, list of films in any specific film genre of Tom Hanks film).

The Semantic Web's Linked Open Data (LOD) cloud interlinks KGs and other data sources using the W3C Resource Description Framework (RDF) [4] and makes accessible on web querying through SPARQL. This LOD cloud is growing rapidly. At the time of this writing, the LOD cloud contains 1,234 datasets with 16,136 links[2]. Several hundred data sets on the Web publish RDF links pointing to DBpedia themselves and thus make DBpedia one of the central interlinking hubs in the Linked Open Data (LOD) cloud [ref]. DBpedia ontology forms a subsumption hierarchy consisting of a standard limited (760) set of classes (types), and recent version of DBpedia has been incorporated a large number (570,276) of YAGO types (mostly file-grained types) by linking YAGO types taxonomy.

---

[1] http://dbpedia.org/resource/Tom_Hanks.

[2] https://lod-cloud.net/.

Although couple of approaches, such as SDType [47] a heuristic approach and, SNCN [40] a hierarchical classification approach have been applied on DBpedia for type inference, these approaches are successful in extracting commonly used coarse types, such as *'Person'*, *'Artist'*, *'Movie'*, or *'Actor'*. In DBpedia, a vast amount of entities missing of fine-grained types (depth of four to six in type hierarchy). For instance, (at the time of this writing) 18805 number of entities listed as American Film class (fine-grained type) within 94996 number of entities from movie class (type) in DBpedia [footnote reference]. However, according to current DBpedia online, only 83 entities (from actor class type) as identified as American Film Actor which evident that 98% of entities missing of this fine-grained entity type.

In recent years, representation learning in form of latent variable methods [14,23,27,31,32,37–39,41,43,53] have increasingly been gained attention for the statistical modeling of KGs, learning latent embeddings for entities and relation-types from the data that can then be used as representations of their semantics. These models have successfully been applied on FB15K [14] dataset, is a subset of Freebase KG which has been commonly used to evaluate various KG completion task, and showing promising results in tasks related to link predictions. DBpedia data are represented in the form of RDF [4] triples <*subject, predicate, object*>, where the subject and object are entities and the predicate is the relation type. The representation learning from DBpedia (large-scale relational data) has therefore become emerged especially for fine-grained entity type inference.

Modeling and fatorizing entire DBpedia is not a trivial task, as DBpedia is very large scale (with millions of entities with billions of facts), and contains heterogeneous information where mappings are created via a world-wide crowd-sourcing effort to extract contents from the information created in various Wikimedia projects. Such information includes infobox templates, categorisation information, images, geo-coordinates, links to external web pages, disambiguation pages, redirects between pages, and links across different language editions of Wikipedia. Besides, A large number of fine-grained types (sub-class) from YAGO type taxonomy are not systematically consistent in the DBpedia ontology. Furthermore, (in the depth of five and six) are not coherently defined in context to sub-class types hierarchy. In addition, A good number of types redundant in DBpedia, such as 5 (five) different types exprese as Actor type class (in Table 3). Although many of these types mapped to DBpedia types using owl:equivalentClass, this leads inconsistency and miss proper fine-grained typing of entities in DBPedia. In this paper, we focus on the extraction of entity fine-grained types, i.e., assigning fine-grained types to – or typing – entities in DBpedia. The major two folds contributions of this paper are as follows:

1. This paper models entire DBpedia with a approach based on a tensor model that learns latent embeddings for entities, relation-types and properties to automatically identify entities to be semantically interpretable by having fine-grained types for connecting them to DBpedia classes. The key idea behind of modelling and applying factorization method is that it uses three-dimensional arrays (tensor) to represent DBpedia and obtain probabilistic likelihoods of

type-relations existing between entities (objects) by applying tensor factorization (TF) techniques on DBpedia.

2. This paper proposes TypePathSample algorithm an efficient way to reduce the computer complexity for the large-size of the dataset, yet operate on a representative subset there of is to use KG partition. This will capture (observing) rich interactions of all the entities of fine-grained types in populating tensor according to fine-grained type entity constraint. This will transform as unobserve from observing interactions of all the entities of coarse-grained types according to fine-grained type entity constraint. Applying this algorithm to DBpedia, we generate multiple samples of the coupled data with domain and type, we fit a Coupled Matrix and Tensor Factorization (CMTF) model to each sample and propose to simultaneous factorization by parallelization.

3. We demonstrate the benefits of this task in the context of fine-grained entity type inference with experiments on a large-scale KG by producing 1.3×10^5 of resources in different fine-grained entity types for person entities from one sample in DBpedia.

This paper is structured as follows. The next section contains related work. In Sect. 2 explain BDpedia Knowledge Graph; modeling and factorizing DBpedia with tensor factorization model. In Sect. 3 In Sect. 3.1 we introduce our approach. In Sect. 3 we describe our experiments. We conclude in Sect. 5.

2   DBpedia Knowledge Graph

DBpedia [1,10,36], a large-scale KG extracted from Wikipedia currently describes 6.6M entities, and 5.5M resources are classified in a consistent ontology, such as 1.5M persons, 840K places, 496K works. Altogether the DBpedia 2016-10 release (see footnote 2) consists of 13 billion pieces of information (RDF triples). Each resource in the DBpedia data set is denoted by a de-referenceable Internationalized Resource Identifier (IRI)- or the Uniform Resource Identifier (URI)-based reference of the form http://dbpedia.org/resource/Name. URI uniquely identifying each entity in Semantic Web KGs. For instance, en entity *Tom Hanks* can be found in DBpedia[3], and in Wikipedia[4]. Every DBpedia entity name resolves to a description-oriented Web document (or Web resource).

DBpedia is served on the web in three forms: First, it is provided in the form of downloadable data sets where each data set contains the results of one of the extractors; second, DBpedia is served via a public SPARQL endpoint and, third, it provides dereferencable URIs according to the Linked Data principles. DBpedia datasets in N3/TURTLE serialisation, and each triple is represented as the form <*head entity, relation, tail entity*>.

The DBpedia data set can be accessed online via a SPARQL query endpoint[5] and as Linked Data[6]. All list of types (coarse-grained or fine-grained) of an

---

[3] http://dbpedia.org/resource/Name.

[4] http://en.wikipedia.org/wiki/Name.

[5] https://dbpedia.org/sparql.

[6] http://mappings.dbpedia.org/server/ontology/classes/.

Table 1. Different RDF triple relations in DBpedia

a. Example of Type relation in a RDF triple	
Subject/head entity	<dbpedia.org/resource/Tom_Hanks>
Predicate/relation	<www.w3.org/1999/02/22-rdf-syntax-ns#type>
Object/tail entity	<dbpedia.org/ontology/Actor>
b. Example of SPARQL query for finding types of an entity	
select distinct ?Subject where ?Subject ?Predicate ?Object filter (?Object = <dbpedia.org/ontology/Actor> && ?Predicate = <www.w3.org/1999/02/22-rdf-syntax-ns#type>)	

entity, or all list of entities for any types can be obtained by SPARQL search on DBpedia (see in Table 1(b));

2.1 Modeling DBpedia

From modelling perspective, tensor representations are appealing to KG because they provide an elegant way to represent multiple RDF triples. The interpretation of DBpedia can be interpreted as a tensor, where first mode of a tensor therefore models the occurrences of all entities as a subject, the second mode models the occurrences of all entities as an object, and the third mode models the different relations, as illustrated in Fig. 2. Entities in DBpedia can be subjects or objects in multiple relations (RDF triples) depending on relation types. For instance, in a relation < *Tom Hanks, starring, Inferno*>, and in another relation <*Inferno, rdf-type, Movie*>, where entity *Inferno* is a subject in one relation and an object in another relation.

The DBpedia ontology consists of 760 classes (such as Thing, Person or Movie) which form a subsumption hierarchy. Figure 1 depicts a part of the DBpedia ontology, indicating the relations from the top class of the DBpedia ontology, i.e. the classes with the highest number of instances. The complete DBpedia ontology can be browsed online (see footnote 3). The file Instance Types (see footnote 4) contains triples of the form <object> <rdf:type> <class> from the mapping based extraction. We can therefore easily model a class as an object in a rdf triple and populate to a tensor. However, a large number of fine-grained types (sub-class) from YAGO type taxonomy are not systematically consistent in the DBpedia ontology. Furthermore, (in the depth of five and six) are not coherently defined in context to sub-class types hierarchy. To address this issue we extended DBpedia type class hierarchy with YAGO type classes for each of DBpedia ontology class using SPARQL [9] query with <http://www.w3.org/ 2000/01/rdf-schema#subClassOf>. For instance, all 2502 [footnote reference] sub classes of DBpedia ontology class for Actor can be derived by this SPARQL query.

Since DBpedia data are high-dimensional but very sparse, we approach the problem of learning positive examples only from DBpedia, by assuming that missing triples are very likely not true. We use the *weighted-tensor* interpretation scheme to effectively model DBPedia for constructing the tensor. The weighting tensor with different values for KG and text data is a particularly important component of our model. Without applying weight in tensor construction, the objective function would place equal importance on fitting both observed and unobserved values. Since DBPedia KB is very large scale, and constructed tensors are therefore often very sparse, this will result in fitting a large number of unobserved data and uncertainty in the observations. The weighting tensor prevents this from happening by emphasizing only the observed (or certain) entries. we approach the problem of learning positive examples only from DBpedia, by assuming that missing triples are very likely not true. DBpedia consisting of e entities and r different relations can then be represented in form of a tensor $X \in R^{e \times e \times r}$ with entities

$$X_{i,j,k} = \begin{cases} 1, & \text{if the relationship } r_k(e_i; e_j) \text{ exists in DBpedia.} \\ 0, & \text{otherwise.} \end{cases} \tag{1}$$

The values ('2' or '1') and ('0') of X_{ijk} come from tensor model are regarded as observed and un-observed data respectively, the representations of DBpedia are therefore becomes possible for tensor factorization.

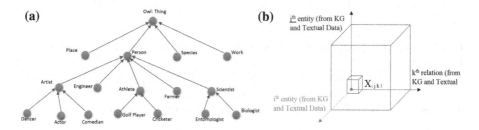

Fig. 1. (a) A part of DBPedia type hierarchy. (b) Representation of a third-order tensor with RDF triples.

The schema information for RDF, which provides the concepts rdfs:domain and rdfs:range for a semantic description of the entities contained in the KG. These concepts are used to represent type-constraints on relation-types by defining the classes or types of entities which they should relate, where the domain covers the subject entity classes and the range the object entity classes in a RDF-Triple. DBPedia domain information can be found in the property of a relation by SPARQL [5] query with http://www.w3.org/1999/02/22-rdf-syntax-ns# Property. For example, a set of actor typed entities can be a set of author typed where these entities are involved with different domains. Ignoring these information (inter-domains collaboration activities of entities) may effect on latent features learning by factorization.

Leveraging Domain Knowledge. Most KGs (such as DBPedia, Freebase, or Yago) store facts about real-world objects covering only numbers of specific domains (e.g. "Movie", "Book", or "Place"). For instance, types in KG such as *actor, film, director,* or *producer* and fine-grained types such as *filmActor, TVActor, regularActor, guestActor, executiveDirector, AssistantProducer* are in *"film"* domain. Given the importance the fine-grained inference task in KG, typed entities (objects) for given fine-grained types in one domain (such as *"film"* domain) are less likely to be entities in other domains (such as *"book"*, or *"place"*). For instance, inferring entities for fine-grained types (such as regularActor, guestActor) would be a typed in Actor, those entities generally are in same domain in KG.

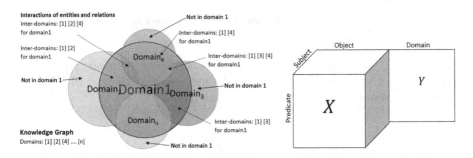

Fig. 2. (left) Representation of a KG with different domains. (right) Modelling domain knowledge in a tensor.

The collaborative activities between the entities in *"film"* domain are therefore higher importance for fine-grained type inference in this domain.

Partitioning via Type Path Hierarchy. Introduce *Fine-grained type entity constraint* for Knowledge Graph: This fine-grained type entity constraint will distinguish and separate types in KG into two sets of types – (a) Coarse-grained types and (b) Fine-grained types. Developing an effective algorithm for type-class path partitioning is an efficient way to reduce the computer complexity for the large-size of the dataset, yet operate on a representative subset thereof is to use KG partition. This will capture (observing) rich interactions of all the entities of fine-grained types in populating tensor according to 'fine-grained type entity constraint'. This will transform as unobserved from observing interactions of all the entities of coarse-grained types according to 'fine-grained type entity constraint'.

Proposed Algorithm:
Input: *Knowledge Graph $G = (T, E, R)$; where set T is a set of 5th level Types in KG, $T = \{t_1, ..., t_{|t|}\}$; set E is a set of entities (objects), $E = \{e_1, ..., e_{|e|}\}$, and E_s as subject set of entities which occur as subject in relation links, where $E_s \in E$; and and E_o as object set of entities which occur as object in relation links where $\{E_o \in E\}$ and, the set R is a set of relations (Predicates) between entities, $R = \{r_1, ..., r_{|r|}\}$.*
Output: set ∇T_c is a list of triples $<< (E_s)_i, (R)_i, (E_o)_i >>$ for each $d_i \in T$ (Types)

Start
01: **function** TYPEPATHSAMPLE (T, E, R)
02:     $d \leftarrow type$
03:     **for each** $r_i \in R$ **do**
04:         **for each** $r_i \in d$ **do**
05:             $source\ (e_s) : R \rightarrow E_s$ ▷ return source(e)
06:             $target\ (e_o) : R \rightarrow E_o$ ▷ return target(e)
07:             $e_d \leftarrow (e_s \cup e_o)$
08:             **while** $e_i \in e_d$ OR $r_i \in d$ **do**
09:                 $T_d \leftarrow$ SELECT-TRIPLES $\{(e_s)_i, (r)_i, (e_o)_i\}$
10:             **end while**
11:                 $\nabla T_c \leftarrow T_c$
12:         **end for**
13:     **end for**
14: **end function**

Applying this algorithm to DBpedia, we generate multiple partitions of data samples of the coupled data with domain and type, we fit a CMTF model to each sample and propose to simultaneous factorization by parallelization.

2.2   Factorizing DBpedia

Classical *Tensor Factorization Models (TFM)* such as *Singular Value Decomposition (SVD)* [20,28], *CANDECOMP/PARAFAC Decomposition (CPD)* [16,30] can be regarded as *Latent Factor Models (LFM)* for multi-relational data [34,43]. Since DBpedia data is multi-relational, the tensor entries from them can be therefore factorized in order to directly comparable by transforming subject entities, relations, object entities and domains to the same latent factor space. The global dependencies are captured during learning the latent representations of each of these dimensions of tensor. The latent ternary correlation subject, object, predicate and domain can be inferred after factorizing the tensor model. We use CP based *Coupled Matrix and Tensor Factorization (CMTF)* [8,9] for deriving the latent relationships between dimensions of the tensor model. After latent factors generation via tensor factorization, we therefore follow *tensor reconstruction* process to reveal new entries that are inferred from the latent factors.

As illustrated Fig. 2, the *CMTF* model, CP factorizes tensor $\mathbf{X} \in \mathbb{R}^{S \times O \times P}$, and a matrix $\mathbf{Y} \in \mathbb{R}^{P \times D}$, can be formulated as

$$f(\mathbf{A}, \mathbf{B}, \mathbf{C}, \mathbf{V}) = \|\mathcal{X} - \|\mathbf{A}, \mathbf{B}, \mathbf{C}\|\|^2 + \|\mathbf{Y} - \mathbf{A}\mathbf{V}^{\mathbf{T}}\| \tag{2}$$

where $\mathcal{X}$ is factorized using a CP model on each mode-n matricization and results in four latent factors matrices, $\mathbf{A} \in \mathbb{R}^{S \times R}$, $\mathbf{B} \in \mathbb{R}^{O \times R}$ and $\mathbf{C} \in \mathbb{R}^{P \times R}$ corresponding to the each dimensions of tensor $\mathcal{X}$, $\mathbf{V} \in \mathbb{R}^{D \times R}$ are the factor matrices extracted from matrix Y through matrix factorization.

Probabilistic Inference. Since DBPedia is multi-relational data, the similarity of entities is therefore determined by the similarity of their relationships, following the intuition that "if two objects are in the same relation to the same object, this is evidence that they may be the same object". The collaborative activities of entities as subjects $\mathbf{A_s} \in \mathbb{R}^{S \times P}$ and objects $\mathbf{A_o} \in \mathbb{R}^{O \times P}$ in relations in a domain can be modelled by the entity matrix $\hat{\mathbf{A}}$, where $\hat{\mathbf{A}}$, is QR matrix factorization [17,29] of $\sum(A_s + A_o)$. For each domain the latent space $\hat{\mathbf{A}}$ therefore reflects the similarity of entities in the relational domain. The type or fine-grained type classes set $C_e = \{t_1, t_2, t_3, ...t_n\}$ where C_e is a set of Types in one KG. A list of type or fine-grained type classes that are considered for given fine-grained type. For each fine-grained type in C_e the candidate entities set, $\hat{E}_t = \{\hat{e}_1, \hat{e}_2, \hat{e}_3, ...\hat{e}_n\}$ where E_t is a set of typed entities in one KG.

We use the Bayes' theorem [24,55] for predicting the class candidate entity E_t that have the highest posterior probability given C_e, $p(C_e|E_t)$. The posterior probability is utilized to calculate the preference probability of an entity e to be fine-grained typed t in C_e type classes by observing current type classes of entity e, and latent similarity of entity e to fine-grained typed entity. The conditional probability can be formulated as:

$$p(C_e|E_t) = \frac{p(E_t|C_e)p(C_e)}{p(E_t)} \tag{3}$$

where *prior probability* $p(C_e)$ is the prior distributions of parameter set C_e in a single domain before E_t is observed, that is relative frequency with which observations from that class occur in a population. Generally, prior probability for fine-grained type classes are lower compared to top level type classes in KG. $p(C_e|E_t)$ is the joint probability of observing type class preference set C_e given E_t, and entity similarity preference given fine-grained type t. Using the assumption of multinomial event model distribution for the Naive Bayes classifier, the posterior probability p_{e_n, t_r} for fine-grained type t_e with fine-grained type class C_e for candidate entity e_n, an instance of E_t, is obtained by multiplying the *prior probability* of t_e, $P(C_e = t_e)$, with the probability of preference candidate entity e_n, an instance of E_t, given t_e, $P(e_n|C_e = t_e)$:

$$p_{e_n, t_e} = p(t_e|C_e) = \sum_{t=1}^{|C_e|} P(e_n|C_e = t_e) \prod_{\hat{e}=1}^{|\hat{E}_e|} P(\hat{E}_e|C_e) \tag{4}$$

where, $P(\hat{E}_e|C_e)$ is probability of likelihood for t_e in C_e, is derived from the entities set, $\hat{E}_t = \{\hat{e}_1, \hat{e}_2, \hat{e}_3, ...\hat{e}_n\}$ where values from reconstructed tensor $\hat{X}$, and entity similarity values from $\hat{A}$ are used.

3   Experiments

3.1   Evaluation and Apply on DBpedia and Freebase

Datasets. We apply and demonstrate the benefits of our task in the context of fine-grained entity type inference applying on on DBpedia 2016–10 release dataset. The DBpedia 2016-10 release dataset[7] published on the year 2017, at the time of this writing this release is latest full version of DBpedia KG. To make ready DBpedia dataset to apply tensor based model, we first A simple java program is used to transfer textual based triples into readable format for applying tensor based model. Prior to apply on DBpedia, we evaluate our approach on Freebase FB15K dataset[8]; FB15K-237 Knowledge Base Completion Dataset[9] and DBpedia 2016-10 release dataset (see Table 2). The FB15K (Bordes et al. 2013), is a subset of Freebase which has been commonly used to evaluate various KG completion models [14,31,32,37,38,53,54]. In the FB15K-237 Knowledge Base Completion Dataset, the triples (entity- textual-entity) are derived from 200 million sentences from the ClueWeb12 corpus coupled with Freebase entity. There are around 3.9 million text descriptions corresponding to the relation types in Freebase. The FB15K-237 dataset has been used in [50,51,56] for embedding representations for textual relations with Freebase entity mention annotations.

Table 2. Datesets used in the experiments.

DBpedia	
Dataset	DBpedia 2016-10 release
# Entities	5.72 million
# Relations as object properties	1,105
# Relations as datatype properties	1,622
# Relations as specialised datatype properties	132
# Entity class types	760
# YAGO class types	570,276
# RDF triples from DBpedia 2016-10 release	494 million
# RDF triples from online DBpedia by SPARQL	1.2 million

Freebase			
Datasets	# Entities	# Relations	# Triples
FB15K	14,951	1,345	486,641
FB15K-327	14,951	2,766,477	3,977,677

Implementation for Experiment. For implementation, we use *tensor-toolbox* [6] and *poblano-toolbox* [2] in Matlab. We construct a 3th order tensor where the tensor size (5.72M × 5.72M × 27K) in 52 different domain with 494M entries from DBpedia. First and second orders of this tensor are defined as *Entity* and third order as *Relation*. We fit tensor factorization based model [41] to this tensor where domain is coupled with relation in tensor; and apply TYPEPATHSAMPLE to make samples of the model. Each sample model is density reduced tensor with same size. For instance, in first sample all other samples tensor entries are transformed to unobserved. For evaluation we apply *weighted tensor scheme* in constructing tensor from Freebase where the tensor size (14951 × 14951 × 2,767,822) with 486541 entries from KG (FB15K); and 4460819 entries from textual dataset (FB15K-237). We then apply *domain-relevance weighted tensor (DrWT)* to construct 4th order tensor with domain entries, where the tensor size

---

[7] https://wiki.dbpedia.org/develop/datasets/dbpedia-version-2016-10.

[8] https://developers.google.com/freebase/.

[9] FB15K-237 Knowledge Base Completion Dataset https://www.microsoft.com/en-us/download/details.aspx?id=5231.

Table 3. Fine-grained Entity Types Inference on DBpedia

Fine-grained types	# Entities present in DBpedia	# Entities new identified
http://dbpedia.org/class/yago/WikicatAmericanFilmActors	83	9787
http://dbpedia.org/class/yago/WikicatTelevisionActors	21	5225
http://dbpedia.org/class/yago/WikicatFilmDirectors	291	5310
http://dbpedia.org/class/yago/WikicatFilmsByAmericanDirectors	101	18805
http://dbpedia.org/class/yago/WikicatFilmProducers	2196	1695
http://dbpedia.org/class/yago/WikicatActionFilms	107	235
http://dbpedia.org/class/yago/WikicatAdventureFilms	110	582
http://dbpedia.org/class/yago/WikicatComedyFilms	124	3524
http://dbpedia.org/class/yago/WikicatHorrorFilms	116	1194
http://dbpedia.org/class/yago/WikicatDramaFilms	189	2360
http://dbpedia.org/class/yago/WikicatCrimeFilms	124	980
http://dbpedia.org/class/yago/WikicatMysteryFilms	118	511
http://dbpedia.org/class/yago/WikicatMusicalFilms	125	374
http://dbpedia.org/class/yago/WikicatFantasyFilms	116	718
http://dbpedia.org/class/yago/WikicatScienceFictionFilms	117	276
http://dbpedia.org/class/yago/WikicatRomanceFilms	109	492
http://dbpedia.org/class/yago/WikicatThrillerFilms	126	560
http://dbpedia.org/class/yago/WikicatAnimatedFilms	104	492
http://dbpedia.org/class/yago/WikicatArtFilmss	192	205
http://dbpedia.org/class/yago/WikicatRomanticComedyFilms	98	256
http://dbpedia.org/class/yago/WikicatShortFilms	204	245
http://dbpedia.org/class/yago/WikicatDocumentaryFilms	109	2105
http://dbpedia.org/class/yago/WikicatWarFilms	114	178
http://dbpedia.org/class/yago/WikicatPoliticalFilms	24	178
http://dbpedia.org/class/yago/WikicatTelevisionFilms	227	1652
http://dbpedia.org/class/yago/WikicatTelevisionActors	5200	248
http://dbpedia.org/class/yago/WikicatAmericanFilmActresses	7278	495
http://dbpedia.org/class/yago/WikicatVoiceActors	708	451
http://dbpedia.org/class/yago/WikicatChildActors	911	98
http://dbpedia.org/class/yago/WikicatMusicalTheatreActors	56	61
http://dbpedia.org/class/yago/WikicatVideoGameActors	17	98
http://dbpedia.org/class/yago/WikicatStageActors	209	2119
http://dbpedia.org/class/yago/WikicatAmericanActors	8823	964

$(14951 \times 14951 \times 2{,}767{,}822 \times 52)$. We use CP [16,30] based *4th-order Tensor Factorization* for the latent factor generation, and use CP-ALS algorithm [19,21, 33] for computing tensor factorization. Since domain information is not depended in one other dimension of the tensor, we use 4th order tensor factorization instead of using *Coupled Matrix Tensor Factorization (CMTF)* [8,9]. We also apply *non-negativity constrain* [35] for effectively interpreting factor components from tensor factorization.

We demonstrate the benefits of our approach in the context of fine-grained entity type inference with experiments on a large-scale KG DBpedia by producing a large number of resources indifferent fine-grained entity types for connecting them to DBpedia type classes. Some new resources unidentified in Film domain in DBpedia are listed in Table 3. In Table 3, new identified entities for fine grained types http://dbpedia.org/class/yago/WikicatAmericanFilmActors http://dbpedia.org/class/yago/WikicatTelevisionActors and http://dbpedia.org/class/yago/WikicatFilmDirectors are 9787; 5225 and 5225 respectfully.

4 Related Work

RESCAL [43] is the state-of-the-art method for link prediction and type inference in KGs that has been used for type inference on YAGO [7] entire KG [44]. This approach defines statistical models for modeling tensor representation of binary relational data on KGs and explains triples via pairwise interactions of latent features Though, YAGO one of the large scale KB in LOD cloud is factorized with RESCAL and able to predict the likelihood of any of the 4.3 x 10^14 possible triples in the YAGO 2 core ontology [44]; DBpedia is not yet modelled with such latent factor model. Paulheim, H. and Bizer, C. proposed SDType algorithm [46,47] a probabilistic method for predicting missing type of entities in DBpedia. Their approach which is based on conditional probabilities, such that predicts approximate types of entities by considering the observed types of subjects and objects in a relation. For each relation, this approach uses the statistical distribution of types in DBpedia based on the property of the subject and object for assuming the types of entities. This approach heuristically suggest that an entity should have certain types if it has certain relations connected to other entities. For example, a statement like <Tom Hanks, starring, Inferno>, this may give result that Tom Hanks is an actor [47].

Though, SDType algorithm has been applied to DBpedia and produced meaningful results in predicting entities for generic (coarse-grained type) classes, (such as actor, writer, or movie); in context to more specific (fine-grained types) classes, (such as American film actor, science-fiction writer, or thriller movie) this heuristic approach is not capable to produce meaningful results. This is because, SDType uses relations between entities as indicators for types, and relations between entities in DBpedia are coherently specific to generic entity types whereas too general to more specific types. Considering previous example, starring relations may be indicators for actor (generic type), however this is too general for all sub-class types of actor (such as film actor, voice actor or television actor) to indicate or distinguish. One recent state-of-the-art fine-grained type entity inference approach [41] which mainly focus on the fine-grained type entity inference task in the KGs via tensor factorization and probabilistic inference methods. First, it looks into the scope of utilizing embedded knowledge inside the KGs that will be efficiently captured to the fine-grained type entity inference task. Besides, it explores the advantages of using linked entity supplementary information to this task by effective incorporation of additional data to KGs. Furthermore, the use of similarity of entities in the KGs is also considered to the fine-grained type entity inference task. Experimental results show that this novel approach has achieved a significant improvement in the accuracy of fine-grained types entity inference in a KG. We models entire DBpedia following this tensor model based approach that learns latent embeddings for entities, relation-types and properties to automatically identify entities to be semantically interpretable by having fine-grained types for connecting them to DBpedia classes.

5   Conclusion

The performance of Web search queries (in case of exploring lists and collections) can be dramatically improved by defining large numbers of these fine-grained entity types in KG. This paper models entire DBpedia with a approach based on a tensor model that learns latent embeddings for entities, relation-types and properties to automatically identify entities to be semantically interpretable by having fine-grained types for connecting them to DBpedia classes. The key idea behind of modelling and applying factorization method is that it uses three-dimensional arrays (tensor) to represent DBpedia and obtain probabilistic likelihoods of type-relations existing between entities (objects) by applying tensor factorization (TF) techniques on DBpedia. This paper proposes a novel way to reduce the computer complexity for the large-size of the dataset, yet operate on a representative subset there of is to use KG partition. Applying this algorithm to DBpedia, we generate multiple samples of the coupled data with domain and type, we fit a Coupled Matrix and Tensor Factorization (CMTF) model to each sample and propose to simultaneous factorization by parallelization. We demonstrate the benefits of this task in the context of fine-grained entity type inference with experiments on a large-scale KG by producing 1.3×10^5 of resources in different fine-grained entity types for person entities from one sample in DBpedia.

References

1. "DBPedia" Public Semantic Knowledge Graph. http://wiki.dbpedia.org/. Accessed 08 Aug 2017
2. "Poblano Toolbox" Poblano - Sandia software - Sandia National Laboratories. https://software.sandia.gov/trac/poblano/. Accessed 29 Aug 2018
3. "Probase" Knowledge Base. https://www.microsoft.com/en-us/research/project/probase/. Accessed 29 Sept 2018
4. "RDF" Resource Description Framework. https://www.w3.org/RDF/. Accessed 29 Sept 2018
5. "SPRQL" Query Language for RDF. https://www.w3.org/TR/rdf-sparql-query/. Accessed 29 Sept 2018
6. "Tensor Toolbox" MATLAB Tensor Toolbox. https://www.sandia.gov/~tgkolda/TensorToolbox/index-2.6.html. Accessed 09 Aug 2018
7. "YAGO" semantic knowledge base. http://www.mpi-inf.mpg.de/departments/databases-and-information-systems/research/yago-naga/yago/. Accessed 09 Aug 2017
8. Acar, E., Kolda, T.G., Dunlavy, D.M.: All-at-once optimization for coupled matrix and tensor factorizations. arXiv preprint arXiv:1105.3422 (2011)
9. Acar, E., Rasmussen, M.A., Savorani, F., Næs, T., Bro, R.: Understanding data fusion within the framework of coupled matrix and tensor factorizations. Chemometr. Intell. Lab. Syst. **129**, 53–63 (2013)

10. Auer, S., Bizer, C., Kobilarov, G., Lehmann, J., Cyganiak, R., Ives, Z.: DBpedia: a nucleus for a web of open data. In: Aberer, K., et al. (eds.) ASWC/ISWC -2007. LNCS, vol. 4825, pp. 722–735. Springer, Heidelberg (2007). https://doi.org/10.1007/978-3-540-76298-0_52

11. Azmy, M., Shi, P., Lin, J., Ilyas, I.: Farewell freebase: migrating the simplequestions dataset to DBpedia. In: Proceedings of the 27th International Conference on Computational Linguistics, pp. 2093–2103 (2018)

12. Berant, J., Chou, A., Frostig, R., Liang, P.: Semantic parsing on freebase from question-answer pairs. In: Proceedings of the 2013 Conference on Empirical Methods in Natural Language Processing, pp. 1533–1544 (2013)

13. Bollacker, K., Evans, C., Paritosh, P., Sturge, T., Taylor, J.: Freebase: a collaboratively created graph database for structuring human knowledge. In: Proceedings of the 2008 ACM SIGMOD International Conference on Management of Data, pp. 1247–1250. ACM (2008)

14. Bordes, A., Usunier, N., Garcia-Duran, A., Weston, J., Yakhnenko, O.: Translating embeddings for modeling multi-relational data. In: Advances in Neural Information Processing Systems, pp. 2787–2795 (2013)

15. Carlson, A., Betteridge, J., Kisiel, B., Settles, B., Hruschka Jr., E.R., Mitchell, T.M.: Toward an architecture for never-ending language learning. In: AAAI, vol. 5, p. 3 (2010)

16. Carroll, J.D., Chang, J.-J.: Analysis of individual differences in multidimensional scaling via an N-way generalization of "Eckart-Young" decomposition. Psychometrika **35**(3), 283–319 (1970)

17. Chan, T.F.: Rank revealing QR factorizations. Linear Algebra Appl. **88**, 67–82 (1987)

18. Chang, L., Zhu, M., Gu, T., Bin, C., Qian, J., Zhang, J.: Knowledge graph embedding by dynamic translation. IEEE Access **5**, 20898–20907 (2017)

19. Cichocki, A., Zdunek, R., Phan, A.H., Amari, S.: Nonnegative Matrix and Tensor Factorizations: Applications to Exploratory Multi-way Data Analysis and Blind Source Separation. Wiley, Hoboken (2009)

20. De Lathauwer, L., De Moor, B., Vandewalle, J.: A multilinear singular value decomposition. SIAM J. Matrix Anal. Appl. **21**(4), 1253–1278 (2000)

21. De Lathauwer, L., Nion, D.: Decompositions of a higher-order tensor in block terms-part III: alternating least squares algorithms. SIAM J. Matrix Anal. Appl. **30**(3), 1067–1083 (2008)

22. Diefenbach, D., Tanon, T., Singh, K., Maret, P.: Question answering benchmarks for Wikidata. In: ISWC 2017 (2017)

23. Ding, L., Pan, R., Finin, T., Joshi, A., Peng, Y., Kolari, P.: Finding and ranking knowledge on the semantic web. In: Gil, Y., Motta, E., Benjamins, V.R., Musen, M.A. (eds.) ISWC 2005. LNCS, vol. 3729, pp. 156–170. Springer, Heidelberg (2005). https://doi.org/10.1007/11574620_14

24. Domingos, P., Pazzani, M.: On the optimality of the simple bayesian classifier under zero-one loss. Mach. Learn. **29**(2–3), 103–130 (1997)

25. Dong, X., et al.: Knowledge vault: a web-scale approach to probabilistic knowledge fusion. In: Proceedings of the 20th ACM SIGKDD International Conference on Knowledge Discovery and Data Mining, pp. 601–610. ACM (2014)

26. Fabian, M.S., Gjergji, K., Gerhard, W., et al.: YAGO: a core of semantic knowledge unifying WordNet and Wikipedia. In: 16th International World Wide Web Conference, WWW, pp. 697–706 (2007)

27. Franz, T., Schultz, A., Sizov, S., Staab, S.: TripleRank: ranking semantic web data by tensor decomposition. In: Bernstein, A., et al. (eds.) ISWC 2009. LNCS, vol. 5823, pp. 213–228. Springer, Heidelberg (2009). https://doi.org/10.1007/978-3-642-04930-9_14

28. Golub, G.H., Reinsch, C.: Singular value decomposition and least squares solutions. Numerische Math. **14**(5), 403–420 (1970)

29. Gu, M., Eisenstat, S.C.: Efficient algorithms for computing a strong rank-revealing QR factorization. SIAM J. Sci. Comput. **17**(4), 848–869 (1996)

30. Harshman, R.A.: Foundations of the PARAFAC procedure: models and conditions for an "explanatory" multimodal factor analysis (1970)

31. He, S., Liu, K., Ji, G., Zhao, J.: Learning to represent knowledge graphs with Gaussian embedding. In: Proceedings of the 24th ACM International on Conference on Information and Knowledge Management, pp. 623–632. ACM (2015)

32. Ji, G., He, S., Xu, L., Liu, K., Zhao, J.: Knowledge graph embedding via dynamic mapping matrix. In: Proceedings of the 53rd Annual Meeting of the Association for Computational Linguistics and the 7th International Joint Conference on Natural Language Processing (Volume 1: Long Papers), vol. 1, pp. 687–696 (2015)

33. Kim, H., Park, H., Eldén, L.: Non-negative tensor factorization based on alternating large-scale non-negativity-constrained least squares. In: Proceedings of the 7th IEEE International Conference on Bioinformatics and Bioengineering, BIBE 2007, pp. 1147–1151. IEEE (2007)

34. Kolda, T.G., Bader, B.W.: Tensor decompositions and applications. SIAM Rev. **51**(3), 455–500 (2009)

35. Lee, D.D., Seung, H.S.: Learning the parts of objects by non-negative matrix factorization. Nature **401**(6755), 788 (1999)

36. Lehmann, J., et al.: DBpedia-a large-scale, multilingual knowledge base extracted from Wikipedia. Semant. Web **6**(2), 167–195 (2015)

37. Lin, Y., Liu, Z., Luan, H., Sun, M., Rao, S., Liu, S.: Modeling relation paths for representation learning of knowledge bases. arXiv preprint arXiv:1506.00379 (2015)

38. Lin, Y., Liu, Z., Sun, M., Liu, Y., Zhu, X.: Learning entity and relation embeddings for knowledge graph completion. In: AAAI, pp. 2181–2187 (2015)

39. Meilicke, C., Fink, M., Wang, Y., Ruffinelli, D., Gemulla, R., Stuckenschmidt, H.: Fine-grained evaluation of rule- and embedding-based systems for knowledge graph completion. In: Vrandečić, D., et al. (eds.) ISWC 2018. LNCS, vol. 11136, pp. 3–20. Springer, Cham (2018). https://doi.org/10.1007/978-3-030-00671-6_1

40. Melo, A., Völker, J., Paulheim, H.: Type prediction in noisy rdf knowledge bases using hierarchical multilabel classification with graph and latent features. Int. J. Artif. Intell. Tools **26**(02), 1760011 (2017)

41. Moniruzzaman, A.B.M., Nayak, R., Tang, M., Balasubramaniam, T.: Fine-grained type inference in knowledge graphs via probabilistic and tensor factorization methods. In: The World Wide Web Conference, pp. 3093–3100. ACM (2019)

42. Nickel, M., Murphy, K., Tresp, V., Gabrilovich, E.: A review of relational machine learning for knowledge graphs. Proc. IEEE **104**(1), 11–33 (2016)

43. Nickel, M., Tresp, V., Kriegel, H.-P.: A three-way model for collective learning on multi-relational data. In: Proceedings of the 28th International Conference on Machine Learning (ICML-11), pp. 809–816 (2011)

44. Nickel, M., Tresp, V., Kriegel, H.-P.: Factorizing YAGO: scalable machine learning for linked data. In: Proceedings of the 21st international conference on World Wide Web, pp. 271–280. ACM (2012)

45. Paulheim, H.: Knowledge graph refinement: a survey of approaches and evaluation methods. Semant. Web **8**(3), 489–508 (2017)

46. Paulheim, Heiko, Bizer, Christian: Type inference on noisy RDF data. In: Alani, H., et al. (eds.) ISWC 2013. LNCS, vol. 8218, pp. 510–525. Springer, Heidelberg (2013). https://doi.org/10.1007/978-3-642-41335-3_32

47. Paulheim, H., Bizer, C.: Improving the quality of linked data using statistical distributions. Int. J. Semant. Web Inf. Syst. (IJSWIS) **10**(2), 63–86 (2014)

48. Porrini, R., Palmonari, M., Cruz, I.F.: Facet annotation using reference knowledge bases. In: Proceedings of the 2018 World Wide Web Conference on World Wide Web, pp. 1215–1224. International World Wide Web Conferences Steering Committee (2018)

49. Suchanek, F.M., Kasneci, G., Weikum, G.: YAGO: a core of semantic knowledge. In: Proceedings of the 16th International Conference on World Wide Web, pp. 697–706. ACM (2007)

50. Toutanova, K., Chen, D.: Observed versus latent features for knowledge base and text inference. In: Proceedings of the 3rd Workshop on Continuous Vector Space Models and their Compositionality, pp. 57–66 (2015)

51. Toutanova, K., Chen, D., Pantel, P., Poon, H., Choudhury, P., Gamon, M.: Representing text for joint embedding of text and knowledge bases. In: Proceedings of the 2015 Conference on Empirical Methods in Natural Language Processing, pp. 1499–1509 (2015)

52. Wang, Q., Mao, Z., Wang, B., Guo, L.: Knowledge graph embedding: a survey of approaches and applications. IEEE Trans. Knowl. Data Eng. **29**(12), 2724–2743 (2017)

53. Wang, Z., Zhang, J., Feng, J., Chen, Z.: Knowledge graph and text jointly embedding. In Proceedings of the 2014 Conference on Empirical Methods in Natural Language Processing (EMNLP), pp. 1591–1601 (2014)

54. Xiao, H., Huang, M., Zhu, X.: TransG: a generative model for knowledge graph embedding. In: Proceedings of the 54th Annual Meeting of the Association for Computational Linguistics (Volume 1: Long Papers), vol. 1, pp. 2316–2325 (2016)

55. Zhang, H.: The optimality of naive bayes. AA **1**(2), 3 (2004)

56. Zhang, J., Lu, C.-T., Cao, B., Chang, Y., Philip, S.Y.: Connecting emerging relationships from news via tensor factorization. In: 2017 IEEE International Conference on Big Data (Big Data), pp. 1223–1232. IEEE (2017)

57. Zupanc, K.: Davis, J.: Estimating rule quality for knowledge base completion with the relationship between coverage assumption. In: Proceedings of the Web Conference 2018, pp. 1–9 (2018)

Author Index

Printed in the United States
By Bookmasters